The Storytelling Animal

"They say we spend multiple hours immersed in stories every day. Very few of us pause to wonder why. Gottschall lays bare this quirk of our species with deft touches, and he finds that our love of stories is its own story, and one of the grandest tales out there—the story of what it means to be human."
  —**Sam Kean, author of** *The Disappearing Spoon: And Other True Tales of Madness, Love, and the History of the World from the Periodic Table of the Elements*

"Insightful . . . Gottschall wears his erudition lightly, displaying a deep knowledge."   —*Boston Globe*

"A lively introduction to a set of questions raised by amateurs and academics about why we like to read, why stories still compel us, and why even everyday experience can, at times, have the feel of epic."   —*San Francisco Chronicle*

"Story is not the icing, it's the cake! Gottschall eloquently tells you 'how come' in his well-researched new book."
  —**Peter Guber, CEO, Mandalay Entertainment and author of the #1** *New York Times* **bestseller,** *Tell to Win*

The Storytelling Animal

The
Storytelling

Animal

{ HOW STORIES MAKE US HUMAN }

Jonathan Gottschall

MARINER BOOKS · HOUGHTON MIFFLIN HARCOURT · BOSTON · NEW YORK

First Mariner Books edition 2013

www.hmhco.com

Library of Congress Cataloging-in-Publication Data
Gottschall, Jonathan.
The storytelling animal: how stories make us human /
Jonathan Gottschall.
p. cm.
Includes bibliographical references and index.
ISBN 978-0-547-39140-3   ISBN 978-0-544-00234-0 (pbk.)
1. Storytelling. 2. Literature and science. I. Title.
GR72.3.G67 2012
808.5'43—dc23
2011042372

Book design by Brian Moore

Printed in the United States of America
DOC 10 9 8 7 6

To Abigail and Annabel, brave Neverlanders

Contents

God made Man because He loves stories.
— ELIE WIESEL, *The Gates of the Forest*

Preface

Statisticians agree that if they could only catch some immortal monkeys, lock them up in a room with a typewriter, and get them to furiously thwack keys for a long, long time, the monkeys would eventually flail out a perfect reproduction of *Hamlet*—with every period and comma and "'sblood" in its proper place. It is important that the monkeys be immortal: statisticians admit that it will take a very long time.

Others are skeptical. In 2003, researchers from Plymouth University in England arranged a pilot test of the so-called infinite monkey theory—"pilot" because we still don't have the troops of deathless supermonkeys or the infinite time horizon required for a decisive test. But these researchers did have an old computer, and they did have six Sulawesi crested macaques. They put the machine in the monkeys' cage and closed the door.

The monkeys stared at the computer. They crowded it, murmuring. They caressed it with their palms. They tried to kill it with rocks. They squatted over the keyboard, tensed, and voided their waste. They picked up the keyboard to see if

it tasted good. It didn't, so they hammered it on the ground and screamed. They began poking keys, slowly at first, then faster. The researchers sat back in their chairs and waited.

A whole week went by, and then another, and still the lazy monkeys had not written *Hamlet,* not even the first scene. But their collaboration had yielded some five pages of text. So the proud researchers folded the pages in a handsome leather binding and posted a copyrighted facsimile of a book called *Notes Towards the Complete Works of Shakespeare* on the Internet. I quote a representative passage:

Sssssssssssssssssssssssssssssssssssssssssssssssssssssssssssssssssssssssssssssnaaaaaa aaa
Aaaaaaaaaaaaaaaaaaaaaaaaaaaaaaaaaaaaasssssssssssssssssssssfsssssfh-gggggggsss
Assfssssssgggggggaaavmlvvssajjjlsssssssssssssssssa

The experiment's most notable discovery was that Sulawesi crested macaques greatly prefer the letter *s* to all other letters in the alphabet, though the full implications of this discovery are not yet known. The zoologist Amy Plowman, the study's lead investigator, concluded soberly, "The work was interesting, but had little scientific value, except to show that 'the infinite monkey theory' is flawed."

In short, it seems that the great dream of every statistician—of one day reading a copy of *Hamlet* handed over by an immortal supermonkey—is just a fantasy.

But perhaps the tribe of statisticians will be consoled by the literary scholar Jiro Tanaka, who points out that although *Hamlet* wasn't technically written by a monkey, it was written by a primate, a great ape to be specific. Sometime in the depths of prehistory, Tanaka writes, "a less than infinite assortment of bipedal hominids split off from a not-quite infinite group of chimp-like australopithecines, and then another quite finite band of less hairy primates split off from the first motley crew of biped. And in a very finite amount of time, [one of] these primates *did* write Hamlet."

And long before any of these primates thought of writing *Hamlet* or Harlequins or Harry Potter stories—long before these primates could envision writing at all—they thronged around hearth fires trading wild lies about brave tricksters and young lovers, selfless heroes and shrewd hunters, sad chiefs and wise crones, the origin of the sun and the stars, the nature of gods and spirits, and all the rest of it.

Tens of thousands of years ago, when the human mind was young and our numbers were few, we were telling one another stories. And now, tens of thousands of years later, when our species teems across the globe, most of us still hew strongly to myths about the origins of things, and we still thrill to an astonishing multitude of fictions on pages, on stages, and on screens—murder stories, sex stories, war stories, conspiracy stories, true stories and false. We are, as a species, addicted to story. Even when the body goes to sleep, the mind stays up all night, telling itself stories.

This book is about the primate *Homo fictus* (fiction man), the great ape with the storytelling mind. You might not realize it, but you are a creature of an imaginative realm called Neverland. Neverland is your home, and before you die, you will spend decades there. If you haven't noticed this before, don't despair: story is for a human as water is for a fish—all-encompassing and not quite palpable. While your body is always fixed at a particular point in space-time, your mind is always free to ramble in lands of make-believe. And it does.

Yet Neverland mostly remains an undiscovered and unmapped country. We do not know why we crave story. We don't know why Neverland exists in the first place. And we don't know exactly how, or even if, our time in Neverland shapes us as individuals and as cultures. In short, nothing so central to the human condition is so incompletely understood.

The idea for this book came to me with a song. I was driving down the highway on a brilliant fall day, cheerfully spinning the FM dial. A country music song came on. My usual response to this sort of catastrophe is to slap frantically at my radio in an effort to make the noise stop. But there was something particularly heartfelt in the singer's voice. So, instead of turning the channel, I listened to a song about a young man asking for his sweetheart's hand in marriage. The girl's father makes the young man wait in the living room, where he stares at pictures of a little girl playing Cinderella, riding a bike, and "running through the sprinkler with a big popsicle grin / Dancing with her dad, looking up at him." The young man suddenly realizes that he is taking something precious from the father: he is stealing Cinderella.

Before the song was over, I was crying so hard that I had to pull off the road. Chuck Wicks's "Stealing Cinderella" captures something universal in the sweet pain of being a father to a daughter and knowing that you won't always be the most important man in her life.

I sat there for a long time feeling sad but also marveling at how quickly Wicks's small, musical story had melted me—a grown man, and not a weeper—into sheer helplessness. How odd it is, I thought, that a story can sneak up on us on a beautiful autumn day, make us laugh or cry, make us amorous or angry, make our skin shrink around our flesh, alter the way we imagine ourselves and our worlds. How bizarre it is that when we experience a story—whether in a book, a film, or a song—we allow ourselves to be invaded by the teller. The story maker penetrates our skulls and seizes control of our brains. Chuck Wicks was in my head—squatting there in the dark, milking glands, kindling neurons.

This book uses insights from biology, psychology, and neuroscience to try to understand what happened to me on that bright fall day. I'm aware that the very idea of bringing science—with its sleek machines, its cold statistics, its unlovely jargon—into Neverland makes many people nervous. Fictions, fantasies, dreams—these are, to the humanistic imagination, a kind of sacred preserve. They are the last bastion of magic. They are the one place where science cannot—should not—penetrate, reducing ancient mysteries to electrochemical storms in the brain or the timeless warfare among selfish genes. The fear is that if you explain the power of Neverland, you may end up explaining it away. As Wordsworth said, you have to murder in order to dissect. But I disagree.

Consider the ending of Cormac McCarthy's novel *The Road*. McCarthy follows a man and his young son as they walk across a dead world, a "scabland," in search of what they most need to survive: food and human community. I finished the novel flopped in a square of sunlight on my living room carpet, the way I often read as a boy. I closed the book and trembled for the man and the boy, and for my own short life, and for my whole proud, dumb species.

At the end of *The Road,* the man is dead, but the boy lives on with a small family of "good guys." The family has a little girl. There is a shard of hope. The boy may yet be a new Adam, and the girl may yet be his Eve. But everything is precarious. The whole ecosystem is dead, and it's not clear whether the people can survive long enough for it to recover. The novel's final paragraph whisks us away from the boy and his new family, and McCarthy takes leave of us with a beautifully ambiguous poem in prose.

> Once there were brook trout in the streams in the mountains. You could see them standing in the amber current where the white edges of their fins wimpled softly in the flow. They smelled of moss in your hand. Polished and muscular and torsional. On their backs were vermiculate patterns that were maps of the world in its becoming. Maps and mazes. Of a thing which could not be put back. Not be made right again. In the deep glens where they lived all things were older than man and they hummed of mystery.

What does that mean? Is it a eulogy for a dead world that will never burgeon again with life, or is it a map of the "world in its becoming"? Might the boy still be alive, out in the living woods with the good guys, fishing trout? Or is the boy gone, slaughtered for meat? No science can answer these questions.

But science *can* help explain why stories like *The Road* have such power over us. *The Storytelling Animal* is about the way explorers from the sciences and humanities are using new tools, new ways of thinking, to open up the vast terra incognita of Neverland. It's about the way that stories—from TV commercials to daydreams to the burlesque spectacle of professional wrestling—saturate our lives. It's about deep patterns in the happy mayhem of children's make-believe and what they reveal about story's prehistoric origins. It's about how fiction subtly shapes our beliefs, behaviors, ethics—how it powerfully modifies culture and history. It's about the ancient riddle of the psychotically creative night stories we call dreams. It's about how a set of brain circuits—usually brilliant, sometimes buffoonish—force narrative structure on the chaos of our lives. It's also about fiction's uncertain present and hopeful future. Above all, it's about the deep mysteriousness of story. Why *are* humans addicted to Neverland? How did we become the storytelling animal?

The Witchery of Story

Lord! When you sell a man a book you don't sell him
just twelve ounces of paper and ink and glue—you
sell him a whole new life. Love and friendship and
humour and ships at sea by night—there's all heaven
and earth in a book, in a real book I mean.

— CHRISTOPHER MORLEY, *Parnassus on Wheels*

HUMAN LIFE IS so bound up in stories that we are
thoroughly desensitized to their weird and witchy
power. So to start this journey, we need to pry
back the veneer of familiarity that keeps us from noticing the
strangeness of story. All you have to do is open up a storybook, almost any storybook, and pay attention to what it does
to you. Take Nathaniel Philbrick's *In the Heart of the Sea.* It's
not a work of fiction, but it's still a storybook, and a wonderful one at that. Philbrick shapes a riveting tale about the
real-life disaster that inspired Herman Melville to write *Moby
Dick:* the sinking of the whaleship *Essex* by a huge and furious
sperm whale.

Before offering you a taste of *In the Heart of the Sea,* I want you to steel yourself. Philbrick is a crafty old wizard; he waves his pen like a wand. The effect is to drag readers' minds out through their eyes, teleporting them across time and halfway around the world. To resist this wizardry, you must concentrate. Don't lose awareness of your chair, or the drone of traffic in the background, or the solid feel of this book in your hands.

Illustration for *Moby Dick* by A. Burnham Shute (1851)

Page one. It is 1821. The whaleship *Dauphin* is zagging off the South American coast. The Nantucket whalemen are straining their eyes for the steamy plumes that announce their quarry. The *Dauphin's* captain, Zimri Coffin, spots a small boat bobbing on the horizon. He roars to the helmsman to bring the boat under his lee. Philbrick writes:

> Under Coffin's watchful eye, the helmsman brought the ship as close as possible to the derelict craft. Even though their momentum quickly swept them past it, the brief seconds during which the ship loomed over the open boat presented a sight that would stay with the crew the rest of their lives.
>
> First they saw bones — human bones — littering the thwarts and floorboards, as if the whaleboat were the seagoing lair of a ferocious man-eating beast. Then they saw the two men. They were curled up in opposite ends of the boat, their skin covered with sores, their eyes bulging from the hollows of their skulls, their beards caked with salt and blood. They were sucking the marrow from the bones of their dead shipmates.

Quick, where were you? Were you still in your chair, noticing the ache in your back and the drone of traffic, the ink printed on this page? Was your peripheral vision picking up your own thumbs on these margins, the patterns on your living room carpet? Or did Philbrick bewitch you? Were you seeing those raw lips working those splintered bones? Those beards full of salt? The bloodstained foam sloshing in the bilge?

To be honest, I gave you a test you couldn't pass. Human minds yield helplessly to the suction of story. No matter how hard we concentrate, no matter how deep we dig in our heels, we just can't resist the gravity of alternate worlds.

Samuel Taylor Coleridge famously declared that experi-

encing a story—any story—requires the reader's "willing suspension of disbelief." In Coleridge's view, a reader reasons
thus: "Yes, I know Coleridge's bit about the Ancient Mariner
is bunk. But in order to enjoy myself, I have to silence my inner skeptic and temporarily believe that the Ancient Mariner
is real. Okay, there! Done!"

But as the Philbrick snip illustrates, *will* has so little do
with it. We come in contact with a storyteller who utters a
magical incantation (for instance, "once upon a time") and
seizes our attention. If the storyteller is skilled, he simply in-
vades us and takes over. There is little we can do to resist,
aside from abruptly clapping the book shut. But even then,
the image of a starving man trying to gnaw some life from a
comrade's bones will linger in our imaginations.

"Bloodstained foam sloshing in the bilge?" Yes, you caught
me in a lie. I invented that detail to make Philbrick's scene
even more vivid and visceral. But I'm not alone. While reading from *In the Heart of the Sea,* your mind told a prodigious
number of lies as well. You were just more careful about com-
mitting them to print.

When you read the Philbrick scene, it came alive in your
mind. Let me ask you, what did Captain Coffin look like?
Was he young or old? Did he wear a tricorn hat or a floppy-
brimmed deal? What color was his coat? What color was his
beard? How many men crowded the *Dauphin*'s deck? What
press of sail was the good ship bearing? Was the day gray or
blue? Was the swell heavy or light? What sort of rags, if any,
were the two shipwrecked cannibals wearing?

Like Tom Sawyer whitewashing the fence, authors trick
readers into doing most of the imaginative work. Reading is
often seen as a passive act: we lie back and let writers pipe

joy into our brains. But this is wrong. When we experience a story, our minds are churning, working hard.

Writers sometimes compare their craft to painting. Each word is a daub of paint. Word by word—brushstroke by brushstroke—the writer creates images that have all the depth and crispness of real life. But a close look at the Philbrick passage shows that writers are merely drawing, not painting. Philbrick gives us expert line drawings with hints on filling them in. Our minds supply most of the information in the scene—most of the color, shading, and texture.

When we read stories, this massive creative effort is going on all the time, chugging away beneath our awareness. We meet a character who is "handsome" with "fierce eyes" and cheekbones "like blades." And from those small cues we build a human being who has not only those eyes (dark or light?) or those cheeks (ruddy or pale?) but also a certain type of nose and mouth. I know from reading *War and Peace* that Princess Lise Bolkonskaya is small and girlishly vivacious, with a "short" upper lip that leaves her front teeth cutely exposed. But the princess exists with a sort of physical actuality in my mind that greatly exceeds the information Tolstoy provides.

I also know from *War and Peace* that when young Petya is killed in battle, Captain Denisov is very sad. But how do I know? Tolstoy never tells me so. He never shows me Denisov's tears. All I see is Denisov walking slowly away from Petya's warm corpse. Denisov puts his hands on a fence. He grips the rails.

The writer is not, then, an all-powerful architect of our reading experience. The writer guides the way we imagine but does not determine it. A film begins with a writer producing a screenplay. But it is the director who brings the screenplay to life, filling in most of the details. So it is with any story. A writer lays down words, but they are inert. They need a catalyst to come to life. The catalyst is the reader's imagination.

LOST IN NEVERLAND

It is obvious that small children are creatures of story. My own daughters are four and seven as I write this book, and their lives are drenched in make-believe. They spend most of their waking hours traipsing happily through Neverland. They are either enjoying stories in their books and videos or creating, in their pretend play, wonder worlds of mommies and babies, princes and princesses, good guys and bad. Story is, for my girls, psychologically compulsory. It is something they seem to need in the way they need bread and love. To bar them from Neverland would be an act of violence.

Annabel and Abigail Gottschall at play.

In these respects, my children aren't special. Children the world over delight in stories and start shaping their own pretend worlds as toddlers. Story is so central to the lives of young children that it comes close to defining their existence. What do little kids do? Mostly they do story.

It's different for grownups, of course. We have work to do. We can't play all day. In James Barrie's play *Peter Pan* (1904), the Darling children adventure in Neverland, but eventually they get homesick and return to the real world. The play suggests that kids have to grow up, and growing up means leaving the pretend space called Neverland behind.

But Peter Pan stays in Neverland. He won't grow up. And in this, we are all more like Peter Pan than we know. We may leave the nursery, with its toy trucks and dress-up clothes, but we never stop pretending. We just change how we do it. Novels, dreams, films, and fantasies are provinces of Neverland.

Londoners browse through the library at Holland House after an air raid during the Blitz. Unlike other leisure activities—such as quilting, gambling, or sports—everyone *does* story in one form or another. We do story even under the worst conditions, even during war.

One of the puzzles this book addresses is not just story's existence—which is strange enough—but story's centrality. Story's role in human life extends far beyond conventional novels or films. Story, and a variety of storylike activities, dominates human life. You might suspect that my enthusiasm leads me to exaggerate—that I am selling you too hard and too early. Maybe so. But let's look at the numbers.

Even in an age of anxiety about the demise of the book, publishing books is still big business. And more juvenile and adult fiction is sold each year than all categories of nonfiction combined. Some of fiction's competitors read like fiction in drag. For instance, the New Journalism that arose in the 1960s—which has strongly influenced nonfiction across genres—was about telling true stories using fictional technique. Similarly, we like biographies partly for the same reason we like novels: they both follow richly characterized lead protagonists through the struggles of their lives. And the most popular form of biography—the memoir—is notorious for the way it plays loose with facts in search of the grip of fiction.

Still, it is lamentable that while more than 50 percent of Americans still read fiction, we don't read a lot of it anymore. According to a 2009 survey by the Bureau of Labor Statistics, the average American reads for just over twenty minutes per day, and that figure includes everything from novels to newspapers.

We read less than we used to. But this isn't because we have forsaken fiction. No, the page has simply been supplanted by the screen. We spend a staggering amount of time watching fiction on screen. According to a number of different surveys, the average American spends several hours each day watching television programs. By the time American children reach adulthood, they will have spent more time in TV

land than anywhere else, including school. And these numbers don't account for the time we spend in movie theaters or
watching DVDs. When you add in these figures, Americans
spend about nineteen hundred hours per year awash in the
glow of television and movie screens. *That's five hours per day.*

Of course, not all of this screen time is spent watching
comedies, dramas, and thrillers. People also watch news, documentaries, sports, and a hybrid genre called "reality TV,"
which may not quite be fiction but is also not nonfiction.
Still, almost all of our time in theaters and in front of DVD
players is story time, and television remains largely a fiction-
delivery technology.

Then there's music. The musicologist and neuroscientist Daniel Levitin estimates that we hear about five hours of
music per day. It sounds impossible, but Levitin is counting
everything: elevator music, movie scores, commercial jingles,
and all the stuff we mainline into our brains through earbuds.

Of course, not all music tells a story. There are also symphonies, fugues, and avant-garde soundscapes blending wind chimes and bunny screams. But the most popular brand of music tells stories about protagonists struggling to get what they want—most often a boy or a girl. Singers might work in meter and rhyme, and alongside guitarists and drummers, but that does not alter the fact that the singer is telling a story—it only disguises it.

So far we have considered fictions created by artists. But what about the stories we tell ourselves? We are at our most creative at night. When we sleep, the untired brain dreams richly, wildly, and at great length. Consciousness is altered in dreams but not extinguished. We just have a limited ability to remember the adventures we consciously experience throughout the night. (People vary in their ability to remember dreams, but sleep lab studies show that virtually everyone dreams.) In dreams, our brains—like cheating spouses—live a whole separate existence that they conceal from the waking mind.

Scientists used to believe that humans dreamed in a vivid and storylike way only during their REM sleep cycles. If this were true, people would spend about two hours per night—and six years per life—spontaneously scripting and screening night stories in the theaters of their minds. It's amazing that even people who are dull by day can be so creative by night. And it's more amazing that dream researchers now know that storylike dreams actually occur independent of REM and across the whole sleep cycle. Some researchers think that we dream almost all night long.

And we don't stop dreaming when we wake. Many, perhaps most, of our waking hours are also spent in dreams. Daydreams are hard to study scientifically, but if you tune in to your stream of consciousness, you will discover that day-dreaming is the mind's default state. We daydream when driving, when walking, when cooking dinner, when getting dressed in the morning, when staring off into space at work. In short, whenever the mind is not absorbed in a mentally de-manding task—say writing a paragraph like this one or do-ing some difficult calculations—it will get restless and skip off into la-la land.

Clever scientific studies involving beepers and diaries sug-gest that an average daydream is about fourteen seconds long and that we have about two thousand of them per day. In other words, we spend about half of our waking hours—one-third of our lives on earth—spinning fantasies. We daydream about the past: things we should have said or done, working through our victories and failures. We daydream about mun-dane stuff, such as imagining different ways of handling a con-flict at work. But we also daydream in a much more intense, storylike way. We screen films with happy endings in our minds, where all our wishes—vain, aggressive, dirty—come true. And we screen little horror films, too, in which our worst fears are realized.

Some people heap scorn on those Walter Mittys who build castles in the air. But the imagination is an awesome mental tool. While our bodies are always locked into a specific here and now, our imaginations free us to roam space-time. Like powerful sorcerers, all humans can see the future—not a clear and determined future, but a murky, probabilistic one.

What will happen, you wonder, if you yield to your pow-erful need to kick your boss in the testicles? You fire up your imagination to find out. You zoom forward in time. You *see* your boss's smug face. You *hear* your shoe sizzling through the air. You *feel* the contact—squishy at first, then hard. And the simulation convinces you that if you kick your boss, he might return the favor. And then fire you (or call the cops). So you keep your foot holstered; you stew in your cubicle.

But our immersion in story goes beyond dreams and fantasies, songs and novels and films. There is much, much more in human life that is thoroughly infiltrated by fiction.

NOT FICTION, BUT FICTIONY

Pro wrestling is closer to ham theater than sport. The spectacle, all choreographed in advance, gives us elaborate story lines with heroes to love and heels to hate: the pompous magnate, the all-American boy, the evil communist, the effeminate narcissist. It gives us all the grandiose pomp and scale—all the fearless bellowing and overacting—of opera. The fake violence of pro wrestling is exciting. But every atomic drop, Mongolian chop, and camel clutch also advances the plot of a slapstick melodrama about who slept with whose wife, who betrayed whom, who really loves America, and who only pretends to.

Real combat sports obey similar storytelling conventions. Boxing promoters have long understood that fights don't attract fan interest (and dollars) unless they feature compelling personalities and backstories. Prefight hype shapes a story about why the men are fighting and, usually, how they came to despise each other. Fight hype is notoriously fictionalized—men who are friendly off camera pretend to hate each other for the sake of good drama. Without a strong backstory, a fight tends toward dullness, no matter how furious the action. It's like watching the climax to a great film without first watching the buildup that gives the climax its tension.

Vince McMahon, CEO of World Wrestling Entertainment (WWE). In re-branding pro wrestling as "sports entertainment," McMahon shattered "kay-fabe"—wrestling's long-standing refusal to admit that its violent stories were fake. McMahon describes a season of pro wrestling as a serial novel whose story lines culminate in the annual WrestleMania extravaganza. In Barry Blaustein's wrestling documentary *Beyond the Mat,* McMahon is asked to describe WWE's product. He smiles impishly and says, "We make movies."

Pro wrestling is pure fiction, but it only exaggerates what we find in legitimate sports broadcasting, where an announcer—a skilled narrative shaper—tries to elevate a game to the level of high drama. Olympic coverage, for example, is thick with saccharine docudramas about the athletes' struggles. So when the starting gun finally sounds, we are able to root for the competitors as struggling heroes in an epic battle—we are able to feel more fully the thrill of victory and the agony of defeat. A recent *New York Times Magazine* article by Katie Baker makes a similar point. Increasing numbers of women watch televised sports precisely because broadcasters have learned to package it as "interpersonal drama." For female fans, Baker argues, the NFL's appeal overlaps with that of *Grey's Anatomy:* "characters, stories, rivalries and Heartbreak." (I'm not sure it's any different for guys. ESPN Radio is clearly targeted at men, but it showcases relatively few actual sporting events. Instead, the network mainstay is dishy, often catty talk shows about sports personalities. Is LeBron James a jerk for jilting Cleveland? Is "Big Ben" Roethlisberger a sex criminal? Will Brett Favre stay retired, and are the penisgate photos a frame-up?)

Storytelling is the spine of televised sports. This struck me powerfully during the 2010 Masters Golf Tournament. It was the great Tiger Woods's first tournament after a long and lurid sex scandal. Even people who dislike golf didn't want to miss this new chapter in the Tiger saga: would the fallen titan struggle to his feet, or would his bad behavior prove karmic? The broadcast's subplot focused on Tiger's main rival, Phil Mickelson, whose mother and wife were both fighting cancer. The announcers shaped a narrative in which every missed putt or crushed drive was intensely meaningful thanks to these bigger stories.

Mickelson won. Striding victorious from the eighteenth green, he stopped to embrace his cancer-stricken wife. The camera caught a single tear rolling down Mickelson's cheek. It wasn't quite a storybook ending. It was a little too shameless for a storybook.

Television shows such as *Law and Order* and *Survivor* give us story. And they are liberally peppered with breaks in which we are given more story. Social scientists define television commercials as "fictional screen media"; they are half-minute short stories.

A commercial rarely just *says* that a laundry detergent works well; it *shows* that it does through a story about an over-worked mom, rascally kids, and a laundry room triumph. ADT Security Services terrifies us into buying home alarms by showing short films where helpless women and children are rescued from wild-eyed home invaders. Jewelry stores get men to buy sparkly little rocks by screening stories in which besotted suitors pinpoint the exact price of a woman's love: two months' salary. Some ad campaigns are designed around recurring characters in multipart stories, such as the humorous "caveman" ads for Geico insurance or the inspired "Messin' with Sasquatch" ads for Jack Link's Beef Jerky. The latter say nothing about the product, by the way. They just tell stories about beef-jerky-loving guys who foolishly harass an innocent Sasquatch and earn a violent comeuppance.

Humans are creatures of story, so story touches nearly every aspect of our lives. Archaeologists dig up clues in the stones and bones and piece them together into a saga about the past. Historians, too, are storytellers. Some argue that many of the accounts in school textbooks, like the standard story of Columbus's discovery of America, are so rife with distortions and omissions that they are closer to myth than history. Business executives are increasingly told that they must be creative storytellers: they have to spin compelling narratives about their products and brands that emotionally transport consumers. Political commentators see a presidential election not only as a contest between charismatic politicians and their ideas but also as a competition between conflicting stories about the nation's past and future. Legal scholars envision a trial as a story contest, too, in which opposing counsels construct narratives of guilt and innocence—wrangling over who is the real protagonist.

A recent article in *The New Yorker* dwells on the role of story in court. The author, Janet Malcolm, describes a sensational murder trial in which a woman and her lover were accused of killing the woman's husband. Malcolm says that the prosecuting attorney, Brad Leventhal, began his opening statement "in the manner of an old-fashioned thriller." Here's Leventhal:

> It was a bright, sunny, clear, brisk fall morning, and on that brisk fall morning, a young man, a young orthodontist by the name of Daniel Malakov, was walking down 64th Road in the Forest Hills section of Queens County just a few miles from where we are right now. With him was his little girl, his four-year-old daughter, Michelle . . . As Daniel stood outside the entrance to Annandale Playground, just feet from the entrance to that park, just feet from where his little girl stood, the defendant Mikhail Mallayev stepped out as if from nowhere. In his hand he had a loaded and operable pistol.

Leventhal won his verdict largely because he was able to craft a better story from the shaky facts of the case than his opposing counsel, who wasn't as gifted a storyteller.

Like Malcolm's piece in *The New Yorker,* much good journalism is shaped in an intensely storylike way—that's part of what we mean when we call journalism "good." Malcolm doesn't provide a neutral account of what happened. Rather, she takes the chaotic events of a murder and a long trial and weaves them into a suspenseful, character-driven narrative with all of the page-turning appeal of fiction. Here is the way she turns Leventhal into a character in her story: "Brad Lev-

Artist's rendering of the big bang. Science, I argue, can help us make sense of storytelling. But some say that science is a grand story (albeit with hypothesis testing) that emerges from our need to make sense of the world. The storylike character of science is most obvious when it deals with origins: of the universe, of life, of storytelling itself. As we move back in time, the links between science's explanatory stories and established facts become fewer and weaker. The scientist's imagination becomes more adventurous and fecund as he or she is forced to infer more and more from less and less.

enthal . . . is an exceptionally formidable trial lawyer. He is a short, plump man with a mustache, who walks with the darting movements of a bantam cock and has a remarkably high voice, almost like a woman's, which at moments of excitement rises to the falsetto of a phonograph record played at the wrong speed."

We tell some of the best stories to ourselves. Scientists have discovered that the memories we use to form our own life stories are boldly fictionalized. And social psychologists point out that when we meet a friend, our conversation mostly consists of an exchange of gossipy stories. We ask our friend "What's up?" or "What's new?" and we begin to narrate our lives to one another, trading tales back and forth over cups of coffee or bottles of beer, unconsciously shaping and embellishing to make the tales hum. And every night, we reconvene with our loved ones at the dinner table to share the small comedies and tragedies of our day.

Then there are the rich stories in the bedrock of all religious traditions. There are the story forms of jokes and urban legends about partying hard in Las Vegas and waking up minus one kidney. And what about poetry or standup comedy or the rapid rise of increasingly storylike video games that allow a player to *be* a character in a virtual reality drama? What about the way many of us serialize our autobiographies in Facebook and Twitter posts?

We'll come back to these varied forms of storytelling later. For now, I think the point is clear: The human imperative to make and consume stories runs even more deeply than literature, dreams, and fantasy. We are soaked to the bone in story.

But why?

THE STORY PEOPLE

To see what a hard question this is, let's conduct a fanciful, but hopefully illuminating, thought experiment. Throw your mind back into the mists of prehistory. Imagine that there are just two human tribes living side by side in some Afri-

!Kung San storyteller, 1947.

can valley. They are competing for the same finite resources:
one tribe will gradually die off, and the other will inherit the
earth. One tribe is called the Practical People and one is called
the Story People. The tribes are equal in every way, except in
the ways indicated by their names.

Most of the Story People's activities make obvious biological sense. They work. They hunt. They gather. They seek out
mates, then jealously guard them. They foster their young.
They make alliances and work their way up dominance hi-
erarchies. Like most hunter-gatherers, they have a surprising

amount of leisure time, which they fill with rest, gossip, and stories—stories that whisk them away and fill them with delight.

Like the Story People, the Practical People work to fill their bellies, win mates, and raise children. But when the Story People go back to the village to concoct crazy lies about fake people and fake events, the Practical People just keep on working. They hunt more. They gather more. They woo more. And when they just can't work anymore, the Practical People don't waste their time on stories: they lie down and rest, restoring their energy for useful activity.

Of course, we know how this story ends. The Story People prevail. The Story People are us. If those strictly practical people ever existed, they don't anymore. But if we hadn't known this from the start, wouldn't most of us have bet on the Practical People outlasting those frivolous Story People?

The fact that they didn't is the riddle of fiction.

The Riddle of Fiction

It seems incredible, the ease with which we sink through books
quite out of sight, pass clamorous pages into soundless dreams.

— WILLIAM GASS, *Fiction and the Figures of Life*

I FACE THE HEAVY security door. I punch my code into
the keypad. The lock clicks, and I step through the door
into the entryway. I smile a greeting at the director doing
paperwork in her office. I sign the visitors log, open an inte-
rior gate, and am inside the asylum that I visit most days after work.

The room is wide and long and high-ceilinged. It has hos-
pital-hard floors and fluorescent lights. Colorful art is taped
to the walls, and safety scissors lie spread-eagle on the tables.
I smell lemony antiseptic and the cafeteria lunches of Tater
Tots and Beefaroni. As I make my way toward the back of the
room, the inmates babble and yell and bawl and snarl. Some
wear ordinary clothes; others are dressed like ninjas, nurses,
or frilly princesses. Many of the males brandish improvised

weapons; many of the females hold magic wands or swaddled infants.

It's disconcerting. The inmates can see things that I can't—and hear, feel, and taste them, too. There are wicked men lurking in the shadows, and monsters, and the salt smell of the ocean, and the mists of the mountains where a lost baby is wailing for her mother.

Small bunches of inmates seem to be sharing the same hallucination. They fight danger or flee from it as one. They cooperate in cooking fake suppers for little babies who just won't behave. As I continue on toward the back corner of the room, one hero warns me that I am about to step into the jaws of the dragon he is slaying. I thank him. The bold fighter asks a question, and as I veer toward safety, I answer, "I'm sorry, buddy, I don't know when your mom will be here."

At the back of the room, two princesses are tucked in a nook made out of bookshelves. The princesses are sitting Indian-style in their finery, murmuring and laughing—but not with each other. They are both cradling babies on their laps and babbling to them, as mothers do. The small one with the yellow hair notices me. Leaping to her feet, she drops her baby on his head. "Daddy!" Annabel cries. She flies to me, and I sweep her into the air.

At about the age of one, something strange and magical buds in a child. It reaches full bloom at the age of three or four and begins to wilt by seven or eight. At one, a baby can hold a banana to her head like a phone or pretend to put a teddy bear to bed. At two, a toddler can cooperate in simple dramas, where the child is the bus driver and the mother is the passenger, or where the father is the child and the child is the father. Two-year-olds also begin learning how to develop a character. When playing the king, they pitch their voices differently than when they are playing the queen or the meowing cat. At three or four, children enter into the golden age of pretend play, and for three or four more years, they will be masters of romps, riots, and revels in the land of make-believe.

Children adore art by nature, not nurture. Around the world, those with access to drawing materials develop skills in regular developmental stages. Children adore music by nature. I remember how my own one-year-olds would stand and "dance" to a tune: smiling toothlessly, bobbing their huge heads, flailing their hands. And by nature children thrill to fictions in puppet shows, TV cartoons, and the storybooks they love to tatters.

To children, though, the best thing in life is play: the exuberance of running and jumping and wrestling and all the danger and splendor of pretend worlds. Children play at story by instinct. Put small children in a room together, and you will see the spontaneous creation of art. Like skilled improv performers, they will agree on a dramatic scenario and then act it out, frequently breaking character to adjust the scenario and trade performance notes.

Children don't need to be tutored in story. We don't need to bribe them to make stories like we bribe them to eat broccoli. For children, make-believe is as automatic and insuppressible as their dreams. Children pretend even when they don't have enough to eat, even when they live in squalor. Children pretended in Auschwitz.

Why are children creatures of story?

To answer this question, we need to ask a broader one first: why do humans tell stories at all? The answer may seem obvious: stories give us joy. But it isn't obvious that stories *should* give us joy, at least not in the way it's biologically obvi-

Impoverished Indonesian children in the garbage dump where they play.

ous that eating or sex should give us joy. It is the joy of story that needs explaining.

The riddle of fiction comes to this: Evolution is ruthlessly utilitarian. How has the seeming luxury of fiction *not* been eliminated from human life?

The riddle is easy to pose but hard to solve. To begin to see why, hold your hand up in front of your face. Rotate it. Make a fist. Wiggle your fingers. Press each fingertip to your thumb, one after another. Pick up a pencil and manipulate it. Tie your shoelaces.

The human hand is a marvel of bioengineering. In a compact space, it packs 27 bones, 27 joints, 123 ligaments, 48 nerves, and 34 muscles. Almost everything about the hand is

Clay bison, Tuc d'Audoubert cave, Ariège, France. The riddle of fiction is part of a bigger biological riddle, the riddle of art. Fifteen thousand years ago in France, a sculptor swam, crawled, and squirmed his way almost a kilometer down into a mountain cave. The sculptor shaped a male bison rearing to mount a cow and then left his creation in the guts of the earth. The clay bison are an excellent illustration of the evolutionary riddle of art. Why do people make and consume art when doing so has real costs in time and energy and no obvious biological payoffs?

for something. The nails are for scratching and picking and prying. The fingerprints, or papillary ridges, are crucial to our fine sense of touch. Even the sweat ducts on our hands are arranged with purpose: they keep our hands moist, which improves the stickiness of our grip. (A dry finger slides, which is why you may lick your finger before turning this page.) But the pride of the hand is the fully opposable thumb. Without thumbs, our hands would be only a marginal improvement over a pirate's hook. Other animals, with their thumbless extremities, can merely paw at the world, or butt and scrape it with their hooves. But because we humans have thumbs, we can seize hold of it and manipulate it to our ends.

Using their hands and faces, humans can be eloquent without words.

Now indulge me by asking yourself what might seem like a stupid question: what is your hand *for?*

Well, a hand is obviously for eating. A hand is for caressing. A hand is for making fists and bludgeoning. A hand is for making tools and wielding them. A hand is lascivious: it is for groping and tickling and teasing. Hands are for making sense: we wave them around to amplify what we are saying. My own hands are for all of the above, but these days they are mostly for thumbing through books and typing.

Our hands are tools, but evolution did not shape them for one single thing. The hand is not the biological equivalent of a hammer or a screwdriver; the hand is a multipurpose tool like a Swiss Army knife—it is *for* many things.

What is true for the hand is true for many other body parts. Eyes are mainly for seeing, but they also help us communicate our emotions. They narrow when we sneer and when we laugh. They water when we are very sad and, strangely enough, when we are very happy. We have lips because we need a hole to take in food and breath. But lips are multipurpose, too. We use them to express affection through kisses. We flex our lips to let people know what's going on inside our skulls: if we are happy, sad, or killing mad. And lips, of course, are also for speaking.

What is true for lips and hands is also true for the brain, and the behaviors driven by it. Take generosity. While evolutionary psychologists debate where humans sit on the continuum between selflessness and selfishness, it is obvious that humans behave generously under many conditions. What is generosity for? It is for a lot of things: enhancing reputation, wooing mates, attracting allies, helping kinsmen, banking favors, and so on. Generosity isn't for any one thing, and it wasn't forged by a single evolutionary force. Likewise the human penchant for story. Fiction might be *for* a lot of things.

Like what?

Some thinkers, following Darwin, argue that the evolutionary source of story is sexual selection, not natural selection. Maybe stories, and other art forms, aren't just *obsessed with* sex; maybe they are ways of *getting* sex by making gaudy, peacocklike displays of our skill, intelligence, and creativity—the quality of our minds. Thumb back a few pages to that image of the !Kung San storyteller on page 19. Look at the young woman sitting to the storyteller's left—very pretty, very rapt. *That's* the idea.

Or maybe stories are a form of cognitive play. For the evolutionary literary scholar Brian Boyd, "a work of art acts like a playground for the mind." Boyd suggests that the free play of art, in all its forms, does the same sort of work for our mental muscles that rough-and-tumble play does for our physical muscles.

Or maybe stories are low-cost sources of information and vicarious experience; maybe, to modify Horace, stories delight *in order* to instruct. Through stories we learn about human culture and psychology, without the potentially staggering costs of having to gain this experience firsthand.

Or maybe story is a form of social glue that brings people together around common values. The novelist John Gardner expresses this idea nicely: "Real art creates myths a society can live by instead of die by." Go back again to the !Kung San storyteller. Look how he has brought his people together, skin against skin, mind against mind.

These and other theories are all plausible, and we'll return to them later. But before doing so, we need to tackle a different possibility: that story may be for nothing at all. At least not in biological terms.

YOUR BRAIN ON DRUGS

The Krel made first contact at a professional football game, easing their flying saucer down on the fifty-yard line. A mouth yawned open in the ship's belly, and a ramp protruded like a tongue. The terrified fans watched as an alien named Flash appeared in the portal and staggered down the ramp. Flash had a white-blond brush cut and ears like small, fleshy trumpets. He wore a red jumpsuit with a bolt of lightning tearing across his chest. Flash hurried down the ramp, saying, "Cocaine. We need cocaine."

In John Kessel's short story "Invaders," the Krel cross the universe just to score coke. The earthlings are confused, so the Krel explain that they have a different sense of the aesthetic. For them, the beauty of the cocaine molecule is simply shattering. Cocaine is the universe's most sublime chemical symphony. The Krel don't *do* coke; they experience it as art.

Toward the end of the story, Flash reclines on trash bags in an alley, sharing a crack pipe with a fellow junkie. The alien makes a confession: that talk about the beauty of the cocaine molecule was high-minded nonsense. The Krel, Flash admits, do coke "for kicks."

And that's the point of Kessel's story. Fiction, like cocaine, is a drug. People may invent high-minded aesthetic (or evolutionary) justifications for their fiction habits, but story is just a drug we use to escape from the boredom and brutality of real life. Why do we go to see a Shakespeare play, or watch a film, or read a novel? Ultimately, from Kessel's point of view, it is not to expand our minds, explore the human condition, or do anything else so noble. We do it for kicks.

Many evolutionary thinkers would agree with Kessel's position. What are stories *for? Nothing.* The brain is not *designed for* story; there are glitches in its design that make it *vulnerable to* story. Stories, in all their variety and splendor, are just lucky accidents of the mind's jury-rigged construction. Story may educate us, deepen us, and give us joy. Story may be one of the things that makes it most worthwhile to be human. But that doesn't mean story has a biological purpose.

Storytelling, in this view, is nothing like the opposable thumb—a structure that helped our ancestors survive and reproduce. In this view, story is more akin to the lines on your palm. No matter what your fortuneteller claims, the lines are not maps of your future. They are side effects of the flexion of the hand.

Let's make this point more concrete with an example. I recently watched the silly and poignant Judd Apatow film

Funny People—a "bromance" about a standup comedian (Adam Sandler) with a terminal illness. I liked it: I laughed, I cried, the whole bit.

Why did I enjoy the film? If fiction is an evolutionary side effect, the answer is simple: because I enjoy funny things and the film was funny. I laughed a lot, and laughing makes people feel good. I liked it also because, as a human, I'm nosy and gossip-hungry. And the film let me spy, unseen, on people living at the extremes. I liked the film because it soaked my brain in the heady chemicals associated with wild sex, fistfights, and aggressive humor, without the risk of earning those chemicals honestly.

Other evolutionary thinkers find this side-effect view deeply unsatisfactory. No way, they insist. If story were just pleasurable frippery, then evolution would have long ago eliminated it as a waste of energy. The fact that story is a human universal is strong evidence of biological purpose. Well, maybe. But is it really so easy for natural selection to target the genes that lead me to waste my time on *Funny People* and *Hamlet*—time that could be spent earning money or procreating or doing any number of other things with obvious evolutionary benefits?

No. Because my strong attraction to fiction is deeply interwoven with my attraction to gossip and sex and the thrill of aggression. In short, it would be difficult to get rid of the evolutionary bathwater of story without also throwing out the baby—without doing violence to psychological tendencies that are clearly functional and important.

If you feel as if your brain is being twisted into a knot, you're not alone. I don't know for sure whether story is an evolutionary adaptation or a side effect, and neither at this point does anyone else. Science consists of repeated rounds of

Behind the scenes on a porn set. Storytelling may also be a simple by-product of having an imagination. Maybe once we evolved a "mental holodeck" for game planning and other practical purposes, we realized we could get cheap thrills by uploading fictions onto it. This would parallel the evolution of the computer: we invented it for utilitarian reasons but soon figured out that we could use it to look at naked people doing naughty things.

conjecture and refutation, and when it comes to this particular question—"Why story?"—we are mainly in a conjectural phase. My own view is that we probably gravitate to story for a number of different evolutionary reasons. There may be elements of storytelling that bear the impression of evolutionary design, like the tweezing grip we can make with our fingers and thumbs. There may be other elements that are evolutionary by-products, like the specific pattern of freckles and hair follicles on the backs of our hands. And there may be elements of story that are highly functional now but were not sculpted by nature for that purpose, such as hands moving over the keys of a piano or a computer.

In chapters to come, we'll explore the evolutionary benefits of story, the way that a penchant for pretend has helped humans function better as individuals and as groups. But before we get to the arguments and evidence, we need to prepare the way by returning to the nursery. The carnage and chaos of children's make-believe provides clues to fiction's function.

THE WORK OF CHILDREN

Grownups have a tendency to remember the land of make-believe as a heavenly, sun-kissed bunny land. But the land of make-believe is less like heaven and more like hell. Children's play is not escapist. It confronts the problems of the human condition head-on. As the teacher and writer Vivian Paley says of pretend play, "Whatever else is going on in this network of melodramas, the themes are vast and wondrous. Images of good and evil, birth and death, parent and child, move in and out of the real and the pretend. There is no small talk. The listener is submerged in philosophical position papers, a virtual recapitulation of life's enigmas."

Pretend play is deadly serious fun. Every day, children enter a world where they must confront dark forces, fleeing and fighting for their lives. I've written some of this book at my kitchen table, with the land of make-believe changing shape around me. One day as I sat at the table, my two daughters were making elaborate pretend preparations to run away from home. Earlier they had played dolls on the back deck and then had run screeching through the yard as sharks tried to eat them. (They managed to harpoon the sharks with sticks.) Later that same day, I took a break to play "lost forest children" with my younger daughter, Annabel. She set the scene: Pretend our parents are dead, she told me, "bited by tigers."

From now on we would live deep in a tiger-infested forest, fending for ourselves.

Children's pretend play is clearly about many things: mommies and babies, monsters and heroes, spaceships and unicorns. And it is also about only one thing: trouble. Sometimes the trouble is routine, as when, playing "house," the howling baby won't take her bottle and the father can't find his good watch. But often the trouble is existential. Here's an unedited sequence of stories that preschoolers made up, on the spot, when a teacher asked, "Will you tell me a story?"

- The monkeys, they went up sky. They fall down. Choo choo train in the sky. I fell down in the sky in the water. I got on my boat and my legs hurt. Daddy fall down from the sky. (Boy, three)
- [Baby] Batman went away from his mommy. Mommy said, "Come back, come back." He was lost and his mommy can't find him. He ran like this to come home [she illustrates with arm movements]. He eat muffins and he sat on his mommy's lap. And then him have a rest. He ran very hard away from his mommy like that. I finished. (Girl, three)
- This is a story about a jungle. Once upon a time there was a jungle. There were lots of animals, but they weren't very nice. A little girl came into the story. She was scared. Then a crocodile came in. The end. (Girl, five)
- Once there was a little dog named Scooby and he got lost in the woods. He didn't know what to do. Velma couldn't find him. No one could find him. (Girl, five)
- The boxing world. In the middle of the morning

everybody gets up, puts on boxing gloves and fights. One of the guys gets socked in the face and he starts bleeding. A duck comes along and says, "give up." (Boy, five)

What do the stories have in common? They are short and choppy. They are all plot. They are marked by a zany creativity: flying choo-choos and talking ducks. And they are bound together by a fat rope of trouble: a father and son plummet from the clouds; baby Batman can't find his mother; a girl is menaced by a crocodile; a little dog wanders lost in the woods; a man is bludgeoned and bloodied.

A different collection of 360 stories told by preschoolers features the same kind of terrors: trains running over puppies and kittens; a naughty girl being sent to jail; a baby bunny playing with fire and burning down his house; a little boy slaughtering his whole family with a bow and arrows; a different boy knocking out people's eyes with a cannon; a hunter shooting and eating three babies; children killing a witch by driving 189 knives into her belly. These stories amply support the play scholar Brian Sutton-Smith, who writes, "The typical actions in orally told stories by young children include being lost, being stolen, being bitten, dying, being stepped on, being angry, calling the police, running away or falling down. In their stories they portray a world of great flux, anarchy, and disaster."

Themes of mortal trouble aren't limited to the arguably artificial stories children invent for psychologists. Trouble also runs through transcripts of spontaneous play recorded in homes and daycares. Take this transcript of a preschool play session recorded by Vivian Paley. Three-year-old Marni is rocking an empty crib, humming to herself and looking at a doll's arm that she can see beneath a pile of dress-up clothes.

Teacher: "Where's the baby, Marni? That crib is very empty."

Marni: "My baby went to someplace. Someone is crying."

(Marni stops rocking the crib and looks around. There is a boy shoveling away at the sand table.)

Marni: "Lamar, did you see my baby?"

Lamar: "Yeah she's in a dark forest. It's dangerous in there. You better let me go. It's down in this hole I'm making."

Marni: "Are you the daddy? Bring me my baby, Lamar. Oh, good for you, you finded her."

Teacher: "Was she in the dark forest?"

Marni: "Where was she, Lamar? Don't tell me in a hole. No, not in a hole, not my baby."

Or consider another play session, in which several children act out a spectacularly convoluted plot involving dynamite and princesses, bad guys and pilfered gold, endangered kitties and bold frog-ninja-dwarfs. The dialogue captures the almost-psychedelic creativity and exuberance of children's play: it reads like a page out of Hunter S. Thompson.

"Pretend you're a frog and you jump into a bad guy but you don't know it."

"Grab 'em!"

"He's stealing kitty!"

"Get him, over there, get him!"

"Blast him, grind him up, he got the gold!"

"Meow, meow, meow."

"Here's your kitty, Snow White."

"Are you the dwarfs? The frog dwarfs?"

"We're the ninja dwarfs. The frog is a ninja. Watch out!
We might have to blow this place up again!"

BOYS AND GIRLS

Vivian Paley is a MacArthur Foundation "genius award" winner who has been writing about her experiences as a preschool and kindergarten teacher for decades. In her small masterpiece of kiddie anthropology, *Boys and Girls: Superheroes in the Doll Corner,* Paley describes a yearlong experiment in the psychology of gender. But Paley didn't set out to run an experiment. Her main goal was just to make her class work better, and for that to happen, she needed the boys to behave. In Paley's classroom, the boys were agents of chaos and entropy. They dominated the block corner, where they constructed battleships, starships, and other engines of war and then deployed them in loud, dire battles. The girls kept to the doll corner, where they decked themselves out in dress-up clothes, took care of their babies, chatted about their boyfriends, and usually managed to lure over a boy or two to play the roles of princes or fathers.

Paley was born in 1929. Her teaching career spanned massive changes in the fabric of American culture, not least of all in the standard gender roles of men and women. Yet over her career, pretend play hardly changed at all. As Paley's career progressed from the 1950s through the 2000s, women moved into the workforce and men took on duties at home. But in

Paley's classroom, the calendar always seemed to be stuck at 1955. The children were precious little embodiments of gender stereotypes.

Paley—a loving teacher and a wonderfully sensitive observer of children—*hated* this. Her career was spent mainly at the University of Chicago Laboratory Schools, where the values of the whole institution aligned squarely with Paley's own liberal leanings. The parents of Paley's students mainly avoided buying their daughters Barbie dolls for fear of encouraging unhealthy body images, and few allowed their boys to play with toy guns.

Paley watched in dismay as gender roles slowly hardened in her classroom. The girls were just so . . . *girlie*. They played dolls; they pined for their princes; they rarely ran or wrestled or shouted; they often told stories about bunnies and magical pink hippos. And the boys were so . . . *boyish*. They sprinted and shouted and happily rioted; they shot the whole room full of imaginary bullet holes and scorched it with bombs. Denied toy guns, the boys fashioned them out of vaguely gun-shaped objects such as crayons, and when teachers confiscated those, the boys still had their fingers.

Worst of all, when the boys played pirates or robbers, they needed what all hard men need most: victims. And what better victims could there be than the girls? The boys were constantly slashing or blasting their way into the doll corner, dealing death and dragging away spoils. This would often drive the girls to tears—not so much because they disliked being shot or robbed, but because the boys were ruining their own fantasies. It is hard to play Cinderella when Darth Vader and his stormtroopers keep crashing the ball.

Paley's book *Boys and Girls* is about the year she spent trying to get her pupils to behave in a more unisex way. And it is a chronicle of spectacular and amusing failure. None of Paley's tricks or bribes or clever manipulations worked. For instance, she tried forcing the boys to play in the doll corner and the girls to play in the block corner. The boys proceeded to turn the doll corner into the cockpit of a starship, and the girls built a house out of blocks and resumed their domestic fantasies.

Paley's experiment culminated in her declaration of surrender to the deep structures of gender. She decided to let the girls be girls. She admits, with real self-reproach, that this wasn't that hard for her: Paley always approved more of the girls' relatively calm and prosocial play. It was harder to let the boys be boys, but she did. "Let the boys be robbers," Paley concluded, "or tough guys in space. It is the natural, universal, and essential play of little boys."

I've been arguing that children's pretend play is relentlessly focused on trouble. And it is. But as Melvin Konner demonstrates in his monumental book *The Evolution of Childhood,* there are reliable sex differences in how boys and girls play that have been found around the world. Dozens of studies across five decades and a multitude of cultures have found essentially what Paley found in her midwestern classroom: boys and girls spontaneously segregate themselves by sex; boys engage in much more rough-and-tumble play; fantasy play is more frequent in girls, more sophisticated, and more focused on pretend parenting; boys are generally more aggressive and less nurturing than girls, with the differences being present and measurable by the seventeenth month of life. The psychologists Dorothy and Jerome Singer sum up this research: "Most of the time we see clear-cut differences in the way children play. Generally, boys are more vigorous in their activities, choosing games of adventure, daring, and conflict, while girls tend to choose games that foster nurturance and affiliation."

The Neverland boys inhabit is very dangerous; the threat of death and destruction is everywhere. Boys' time in Neverland consists largely of fighting that threat or fleeing from it. The Neverland of girls is dangerous, too, but not quite so crowded with hobgoblins and ax murderers, and not as focused on exuberant physical play. The sorts of dilemmas girls face are often less extreme, with a focus on workaday domestic crises.

But it is important to stress that girl play only *seems* untroubled when compared to the mayhem of boy play. Risk and darkness seep into the doll corner as well. For example, Paley recounts how, at first glance, it may seem that the girls are sweetly playing mother and baby. But look closer. First, the baby almost gets fed poison apple juice. Then a bad guy

The role of sex hormones in gender generally, and play behavior specifically, is illuminated by a disorder called congenital adrenal hyperplasia (CAH), which results in females being exposed to abnormally high levels of male sex hormones in utero. Girls with CAH are quite normal in most respects, but "affected girls show more boy-typical play, prefer playing with boys more, and are less interested in marriage, motherhood, doll play and infant care." Girls with CAH enjoy rough-and-tumble as much as boys do, and they prefer "boy" toys such as trucks and guns over "girl" toys such as dolls and dress-up clothes.

tries to steal the baby. Then the baby "gets his bones broken off" and is almost set on fire.

Similarly, Paley recounts an incident where two girls playing Rainbow Brite and her flying pony, Starlite (magical characters from a 1980s animated television series), are having dinner together. Everything is going fine until a bad guy named Lurky appears. The cute little characters, played by two cute little girls, have no choice but to kill Lurky with explosives.

Unlike some of the other subjects of this book—fiction or

dreams—almost no one thinks that children's pretend play is some sort of random accident of human evolution. The pioneering child psychologist Jean Piaget, who thought that the fantasy life of children was "a muddle out of which more adequate and orderly ways of thinking will emerge," is now definitely in the minority. These days, experts in child psychology agree that pretend play is *for* something. It has biological functions. Play is widespread in animals, and all but universal in mammals, especially the smart ones. The most common view of play across species is that it helps youngsters rehearse for adult life. From this perspective, children at play are training their bodies and brains for the challenges of adulthood—they are building social and emotional intelligence. Play is important. Play is the work of children.

Sex differences in children's play reflect the fact that biological evolution is slow, while cultural evolution is fast. Evolution hasn't caught up with the rapid changes in men's and women's lives that have occurred mainly in the past one hundred years. Children's play still seems to be preparing girls for lives beside the hearth and preparing boys for lives of action in the world. This is the basic division of labor—men doing the hunting and fighting and women doing most of the foraging and parenting—that has characterized human life over tens of thousands of years. Anthropologists have never found a culture where, say, women do the lion's share of fighting or men do most of the child care.

Writing this, I feel a little like the narrator in Edgar Allan Poe's "The Black Cat." Before tying a noose and hanging the titular feline from a tree, the narrator first digs out the cat's eye with a jackknife. Confessing his crime, he writes, "I blush, I burn, I shudder, while I pen the damnable atrocity!" The idea that gender has deep biological roots is something almost everyone accepts these days but still avoids saying in polite company. It sounds too much like a limit on human potential, especially on the potential of women to move into positions of cultural equality. But the spectacular changes in women's lives over the past half century—driven largely by the way that cheap and reliable contraception has given women control of their fertility—should allay our fears.

When my daughter Annabel announces her plan to become a princess when she grows up, I squirm. I say, "You know you can be other things, like a doctor." And Annabel replies, "I'll be a princess *and* a doctor. And a mommy." And I smile and say, "Okay."

AND DOWN WILL COME BABY

Where do the blood and tears of children's play come from? It's possible that they come partly from the stories we tell them. In the Grimms' collection of fairy tales, for example, children are menaced by cannibal witches, wolves bolt down personified pigs, mean giants and innocent children meet grisly deaths, Cinderella is orphaned, and the ugly stepsisters slash off chunks of their feet in hopes of cramming them into the tiny glass slipper (and this is before getting their eyes pecked out by birds). And then there's a tale called "How the Children Played Butcher with Each Other," which was published in the first edition of the Grimms' tales. Here is the story entire:

A man once slaughtered a pig while his children were looking on. When they started playing in the afternoon, one child said to the other: "You be the little pig, and I'll be the butcher," whereupon he took an open blade and thrust it into his brother's neck. Their mother, who was upstairs in a room bathing the youngest child in a tub, heard the cries of her other child, quickly ran downstairs, and when she saw what had happened, drew the knife out of the child's neck and, in a rage, thrust it into the heart of the child who had been the butcher. She then rushed back to the house to see what her other child was doing in the tub, but in the meantime it had drowned in the bath. The woman was so horrified that she fell into a state of utter despair, refused to be consoled by the servants, and hanged herself. When her husband returned home from the fields and saw this, he was so distraught that he died shortly thereafter.

The standard nursery rhymes are about as bad: babies fall out of trees "cradle and all," a little boy mutilates a dog,

Image from "The Old Witch," an English fairy tale.

an old woman who lives in a shoe cruelly whips her starving children, blind mice are hacked up with carving knives, Cock Robin is murdered, and Jack smashes his skull. In one collection of familiar nursery rhymes, a critic counted eight murders, two choking deaths, one decapitation, seven cases of severed limbs, four cases of broken bones, and more. And in a different study, researchers found that contemporary children's television programs had about five violent scenes per hour, while read-aloud nursery rhymes had fifty-two.

Although fairy tales for modern children have been sanitized, they are still full of disturbing material. For example, while the stepsisters' bloody mutilation has been scrubbed from the versions of "Cinderella" I have read to my girls, the story still describes something much worse: a girl's loving parents die, and she falls into the hands of people who despise her.

So is the storm and strife of children's make-believe just an echo of the trouble children find in the stories we give them? Is the land of make-believe dangerous because, all around the world, children just happen to be exposed to fictions that crackle with trouble?

That possibility, even if it were true, wouldn't really answer this question; it would just prompt a new one: why are the stories of *Homo sapiens* fixated on trouble?

The answer to that question, I think, provides an important clue to the riddle of fiction.

3

Hell Is Story-Friendly

Like the movie screen depicting a maniac in a hockey
mask carving people up with a chainsaw; like *Hamlet* with
its killings and suicides and fratricides and incestuous
adultery; like all the violence, family strife, and catastrophic
sex in Sophocles or on TV or in the Bible, . . . poems
of loss and death, can please the reader mightily.

— ROBERT PINSKY, *The Handbook of Heartbreak:
101 Poems of Lost Love and Sorrow*

ONCE UPON A TIME, a father and daughter were at
the grocery store. They were walking down the cereal aisle. The father was pushing a cart. The cart's
left front wheel clattered and creaked. The daughter, Lily,
was three years old. She was wearing her favorite dress: it was
flowery and flowing, and it fanned out wonderfully when she
twirled. She held her father's index finger in her left fist. In
her right fist, she held the grocery list in a sweaty wad.

The father stopped in front of the Cheerios. He scratched
his stubbled chin and asked Lily, "What kind of cereal are we
supposed to get again?" Lily released his finger, unwadded the

paper, and smoothed it on the curve of her belly. She squinted at the neat feminine script. She ran her index finger over the list of items as if she were reading the words. "Cheerios," Lily announced. The father let Lily choose the big yellow box herself and push it up and over the side of the cart.

Later the father would remember how the people passed them in the aisle. He would remember the way the women smiled at Lily as they cruised by with their carts, and how they nodded approvingly at him as well. He would remember the pimply stock boy passing by with his mop and his sloshing bucket on wheels. He would remember the way Lily's small hand held his finger, and how the throb of her grip lingered after she let go.

And most of all he would remember the short man with the dark glasses and the red baseball cap tugged low—the way he slouched next to the pyramid display of Pop-Tarts, smiling down at Lily as she passed, showing a wet gleam of incisor.

The father and daughter walked a little farther down the aisle. They stopped. Lily hugged her father's thigh. The father cradled her small head to his leg. He stared at the box of sugary cereal she had thrust into his hand, saying, "Daddy, please!" The father slowly shook his head as he read the ingredients, fascinated. (There was no food in this food, just chemical substances such as trisodium phosphate, Red 40, Blue 1, BHT, and pyridoxine hydrochloride.) The father's eyes moved up to scan the nutritional information, counting grams of sugar and fat.

He never felt Lily let go of his leg, never felt her head slip from beneath his sheltering hand. Still staring at the box in his hand the father said aloud, "I'm sorry, baby. This stuff isn't good for us. Mommy will be mad if we buy it."

Lily was silent. The father turned to her, knowing that she

would be standing there with her arms crossed tight, her chin tucked to her clavicle, and her lips pushed out in a pout. He turned, but Lily wasn't there. He spun slowly on his heel, and still Lily wasn't there.

And neither was the short man with the red cap.

Now imagine the story told differently.

Once upon a time, a father and a daughter went to the supermarket. Toward the end of the cereal aisle, Lily saw the red box with the cartoon bunny. She thrust the box of Trix into her father's hand and hugged his leg as a bribe. The father didn't bother to read the ingredients. He said, "Sorry, honey. This stuff is bad for you. Mommy'll be mad if we buy it."

Lily released her father's leg and whipped her head from beneath his sheltering palm. She stomped her feet, locked out her knees, and tucked her hands defiantly in her armpits. Lily scowled up at her father. He tried to give a stern look in reply, but he was weak, and her charms defeated him. He tossed the Trix into the cart and cracked a conspiratorial smile. "We're not afraid of that ol' mommy, are we?"

"Yeah," Lily said. "We're not 'fraid!"

The father and daughter purchased all the items on their list. They drove home in their minivan. The mother only pretended to be angry about the Trix. The little family lived happily ever after.

MIND THE GAP

Ask yourself which story you would rather live, the first one or the second? The second, obviously. The first is a nightmare. But which story would make a better film or a novel? The answer is equally obvious: the one with the wet-toothed man.

The first story draws us in and infects us with the need to know what happens next: Has the toothy man taken Lily? Or is she just hiding behind the Pop-Tarts display, smothering her giggles with both palms?

There is a yawning canyon between what is desirable in life (an uneventful trip to the grocery story) and what is desirable in fiction (a catastrophic trip). In this gap, I believe, lies an important clue to the evolutionary riddle of fiction.

Fiction is usually seen as escapist entertainment. When I ask my students why they like stories, they don't give the most obvious answer: because stories are pleasurable. They know this would be a superficial response. Of course stories give pleasure, but why?

So my students dig for a deeper cause: stories are pleasurable because they allow us to escape. Life is hard; Neverland is easy. When we watch reruns of *Seinfeld* or read a John Grisham novel, we take a short vacation from the pressures of reality. Life hounds us. We hide from it in fiction.

But it's hard to reconcile the escapist theory of fiction with the deep patterns we find in the art of storytelling. If the escapist theory were true, we'd expect stories to be mainly about pleasurable wish fulfillment. In story worlds, everything would go right and good people would never suffer. Here's a plot summary of the kind of story the average joe would be reading (all stories would be written in the rare second person to help the reader identify with the main character):

You are the shortstop for the New York Yankees. You are the greatest ballplayer in the history of the known universe. This season, you've faced 489 pitches and smacked 489 home runs. You live mainly on fried ice cream, which, in lieu of bowls, you spoon from the smooth bellies of the lingerie models who lounge around your swanky bachelor pad. Despite the huge number of calories you consume, you never add an ounce of fat to your chiseled frame. After you retire from baseball, you are unanimously elected president of the United States, and after bringing about peace on earth, you live to see your face carved into Mount Rushmore.

I'm exaggerating, of course, but you get the point: if fiction offers escape, it is a bizarre sort of escape. Our various fictional worlds are—on the whole—horrorscapes. Fiction may temporarily free us from our troubles, but it does so by ensnaring us in new sets of troubles—in imaginary worlds of struggle and stress and mortal woe.

There is a paradox in fiction that was first noticed by Aristotle in the *Poetics*. We are drawn to fiction because fiction gives us pleasure. But most of what is actually in fiction is deeply unpleasant: threat, death, despair, anxiety, Sturm und Drang. Take a look at the carnage on the fiction bestseller lists—the massacres, murders, and rapes. See the same on popular TV shows. Look at classic literature: Oedipus stabbing out his eyes in disgust; Medea slaughtering her children; Shakespeare's stages strewn with runny corpses. Heavy stuff.

But even the lighter stuff is organized around problems, and readers are riveted by their concern over how it will all turn out: Can Dumb and Dumber overcome their obstacles to win mates who are way out of their league? Will Sam and Diane on *Cheers,* or Jim and Pam on *The Office,* get together? Will the mousy librarian in the newest Harlequin romance tame the studly forest ranger? Will Bella choose the vampire or the werewolf? In short, regardless of genre, if there is no knotty problem, there is no story.

A MIRROR OF LIFE?

Stories of pure wish fulfillment don't tempt us, but what about stories that show us life as it is actually lived? A truly mimetic fictional work might describe an accountant trying to finish an important but crushingly boring project.

> The middle-aged man sat at his desk, poking indifferently at his keyboard. He scratched himself furtively, even though he was alone. He rolled his head on his neck. He stared blearily at his screen. He looked hopefully around the office, seeking some excuse not to work. Something to organize. Something to eat. He spun slowly in his chair. Once. Twice. On the third revolution, he saw himself in a window. He made cannibal faces at his reflection. He fingered the bags under his eyes. Then the man shook his head back and forth, took a long swig of cold, sour coffee, and returned to his screen. He poked some keys and bothered his mouse. Soon he was thinking, Maybe I should check my e-mail again.

Now imagine this passage not as a prelude to a story where *something* happens. (For instance: *The man suddenly saw a strange woman—naked, obese—reflected in the window. She was standing behind him, shaking a knife at his back. Or maybe she was just giving him the finger.*) Imagine this story going on, with nothing of interest happening, for fifteen excruciating chapters.

In fact, writers have experimented with stories like this. So-called hyperrealist fiction does away with the old plot contrivances of traditional fiction. It presents wafers of life as we actually experience it. The crime novelist Elmore Leonard has described his novels as life with all the boring parts snipped out. Hyperrealist fiction pastes them back in.

George Gissing.
from the Lithograph by William Rothenstein.

In the 1891 novel *New Grub Street* by George Gissing (pictured here), a character named Harold Biffen writes a novel called *Mr Bailey, Grocer,* which describes the life of an ordinary grocer in absolutely realistic detail and with zero dramatic shaping. Biffen's novel is "unutterably boring" by design; it is about the dull monotony of a man's life. The novel is a work of art but sheer drudgery to read. Disappointed in love and in art, Biffen ends up poisoning himself.

Hyperrealism is interesting as an experiment, but like most fiction that breaks with the primordial conventions of storytelling, almost no one can actually stand to read it. Hyperrealist fiction is valuable mainly for helping us see what fiction *is* by showing us what it *isn't*. Hyperrealism fails for the same reason that pure wish fulfillment does. Both lack the key ingredient of story: the plot contrivance of trouble.

A UNIVERSAL GRAMMAR

Fiction—from children's make-believe to folktales to modern drama—is about trouble. Aristotle was the first to note this, and it is now a staple in English literature courses and creative writing manuals. Janet Burroway's *Writing Fiction* is adamant on the point: "Conflict is the fundamental element of fiction . . . In life, conflict often carries a negative connotation, yet in fiction, be it comic or tragic, dramatic conflict is fundamental because in literature only trouble is interesting. Only trouble is interesting. This is not so in life." As Charles Baxter puts it in another book about fiction, "Hell is story-friendly."

The idea that stories are about trouble is so commonplace as to verge on cliché. But the familiarity of this fact has numbed us to how strange it is. Here is what it means. Beneath all of the wild surface variety in all the stories that people tell—no matter where, no matter when—there is a common structure. Think of the structure as a bony skeleton that we rarely notice beneath its padding of flesh and colorful garments. This skeleton is somewhat cartilaginous—there is flex in it. But the flex is limited, and the skeleton dictates that stories can be told only in a limited number of ways.

Stories the world over are almost always about people (or personified animals) with problems. The people want something badly—to survive, to win the girl or the boy, to find a lost child. But big obstacles loom between the protagonists and what they want. Just about any story—comic, tragic, romantic—is about a protagonist's efforts to secure, usually at some cost, what he or she desires.

Story = Character + Predicament + Attempted Extrication

This is story's master formula, and it is intensely strange. There are a lot of different ways stories *could* be structured. For example, we have already considered escapist fantasies of pure wish fulfillment. But while characters frequently do live happily ever after in fiction, they must always earn their good fortune by flirting with disaster. The thornier the predicament faced by the hero, the more we like the story.

Most people think of fiction as a wildly creative art form. But this just shows how much creativity is possible inside a

The idea that stories slavishly obey deep structural patterns seems at first vaguely depressing. But it shouldn't be. Think of the human face. The fact that all faces are very much alike doesn't make the face boring or mean that particular faces can't startle us with their beauty or distinctiveness. As William James once wrote, "There is very little difference between one man and another; but what little there is, is very important." The same is true of stories.

prison. Almost all story makers work within the tight con-
fines of problem structure, whether knowingly or not. They
write stories around a pattern of complication, crisis, and
resolution.

Over the past one hundred years, some authors, writhing
in their chains, have tried to break free from the prison of
problem structure. The modernist movement in literature was
born when writers realized, much to their horror, that they
were working inside well-established walls of convention and
formula. They sought to take something as old as humanity—the storytelling urge—and "make it new."

Modernist attempts to transcend conventional story were
nothing short of heroic (in the way of doomed but noble re-
bellions). Here's one breathtaking passage from James Joyce's
Finnegans Wake that pretty much gives the flavor of the whole
book:

> Margaritomancy! Hyacinthinous pervinciveness! Flowers. A
> cloud. But Bruto and Cassio are ware only of trifid tongues
> the whispered wilfulness, ('tis demonal!) and shadows shad-
> ows multiplicating (il folsoletto nel falsoletto col fazzolotto dal
> fuzzolezzo), totients quotients, they tackle their quarrel. Sicka-
> moor's so woful sally. Ancient's aerger. And eachway bothwise
> glory signs. What if she love Sieger less though she leave Ruhm
> moan? That's how our oxyggent has gotten ahold of half their
> world. Moving about in the free of the air and mixing with the
> ruck. Enten eller, either or.

In contrast to the conventional, if virtuosic, storytelling
of Joyce's *Dubliners, Finnegans Wake* is almost impossible to
love. It is easy to worship the genius of it—the awesome creativity of the language, the half-loony commitment it took to

write this way for almost seven hundred pages and seventeen years. You can celebrate *Finnegans Wake* as an act of artistic revolt, but you can't enjoy it as a story that takes you out of yourself and infects you with the need to know what happens next.

Gertrude Stein praised herself, along with writers like Joyce and Marcel Proust, for writing fiction in which "nothing much happens . . . For our purposes, events have no importance." Nothing much happens, and aside from English professors, no one much wants to read them. Yes, experimental fictions like *Finnegans Wake* are still in print, but they are mainly sold either to cultured autodidacts dutifully grinding their way through the literary canon, or to college students who are forced to pretend that they have read them.

As the linguist Noam Chomsky showed, all human languages share some basic structural similarities — a universal grammar. So too, I argue, with story. No matter how far we travel back into literary history, and no matter how deep we plunge into the jungles and badlands of world folklore, we always find the same astonishing thing: *their stories are just like ours.* There is a universal grammar in world fiction, a deep pattern of heroes confronting trouble and struggling to overcome.

But there is more to this grammar than the similarities in skeletal structure; there are also similarities in the flesh. As many scholars of world literature have noted, stories revolve around a handful of master themes. Stories universally focus on the great predicaments of the human condition. Stories are about sex and love. They are about the fear of death and the challenges of life. And they are about power: the desire to wield influence and to escape subjugation. Stories are not about going to the bathroom, driving to work, eating lunch, having the flu, or making coffee—*unless those activities can be tied back to the great predicaments.*

Why do stories cluster around a few big themes, and why do they hew so closely to problem structure? Why are stories *this* way instead of all the other ways they could be? I think that problem structure reveals a major function of storytelling. It suggests that the human mind was shaped *for* story, so that it could be shaped *by* story.

THE HERO DIES IN OUR STEAD

Navy fighter pilots have many difficult jobs. But perhaps the greatest challenge they face is landing a fifty-thousand-pound airplane—laden with jet fuel and high explosives—on a five-hundred-foot runway that is skimming across the ocean at up to thirty knots. The aircraft carrier is immense and powerful, but the ocean is more so, and the whole runway moves with the swell. The carrier deck is speckled with people and planes. The belly of the huge ship holds thousands of souls, a terrible array of missiles and bombs, and a nuclear reactor. Navy pilots have to land on this thread of concrete in all kinds of weather and in the black of night. They have to do so without wrecking their planes, killing their shipmates, or causing a nuclear disaster. So before letting young aviators attempt actual landings, instructors strap them into flight simulators that provide much of the benefit of practicing landings, without the potential carnage and hellfire of the real thing.

Landing a jet on an aircraft carrier is complicated. But navigating the intricacies of human social life is more so, and

A Helldiver aircraft circling for landing on the USS *Yorktown* during the Second World War.

the consequences of failure can be almost as dramatic. Whenever people come together in groups, they will potentially mate with one another, befriend one another, or fight one another.

Practice is important. People practice playing basketball or violin in low-stakes environments so that they will perform well—in the stadium or concert hall—when the stakes are high. According to evolutionary thinkers such as Brian Boyd, Steven Pinker, and Michelle Scalise Sugiyama, story is where people go to practice the key skills of human social life.

This argument is not new to evolutionary psychology. It is a variation on one of the traditional justifications for fiction. For example, Janet Burroway argues that low-cost vicarious experience—especially emotional experience—is the primary benefit of fiction. As she puts it, *"Literature offers feelings*

The HBO sitcom *Curb Your Enthusiasm* offers a master's course in the dangerous involutions of social existence. In most episodes, the Aspergerish main character, Larry David (shown here), commits a truly hideous faux pas by failing to understand and negotiate the weird ballet of human interactions.

for which we don't have to pay. It allows us to love, condemn, condone, hope, dread, and hate without any of the risks those feelings ordinarily involve."

The psychologist and novelist Keith Oatley calls stories the flight simulators of human social life. Just as flight simulators allow pilots to train safely, stories safely train us for the big challenges of the social world. Like a flight simulator, fiction projects us into intense simulations of problems that run parallel to those we face in reality. And like a flight simulator, the main virtue of fiction is that we have a rich experience and don't die at the end. We get to simulate what it would be like to confront a dangerous man or seduce some-one's spouse, for instance, and the hero of the story dies in our stead.

So, this line of reasoning goes, we seek story because we enjoy it. But nature designed us to enjoy stories so we would get the benefit of practice. Fiction is an ancient virtual reality technology that specializes in simulating human problems. Interesting theory. Is there any evidence for it beyond problem structure?

TO SIMULATE IS TO DO

The television show I'm watching cuts to a commercial for the National Football League. The ad absorbs my attention before I can think of surfing away. A dark-skinned boy is racing in slow motion over green grass straight into my living room. The boy is beaming, tossing glances—semi-scared, semi-delighted—at someone or something to his right. Suddenly a huge, handsome young man (a defensive lineman for the Houston Texans) sweeps in from outside the frame. He scoops the giggling boy up like a football and runs at the camera, still in slow motion. The man and the boy are both smiling to crack their faces. Sitting alone in my living room, I cannot help but smile so broadly it stings.

In the 1990s, quite by accident, Italian neuroscientists discovered mirror neurons. They implanted electrodes into a monkey's brain to discover which neural regions were responsible for, say, commanding the hand to reach out and grab a nut. In short, the scientists discovered that specific areas of a monkey's brain light up not only when they grab a nut, but whenever they see another monkey or person grab a nut.

There has since been a flood of mirror neuron research in

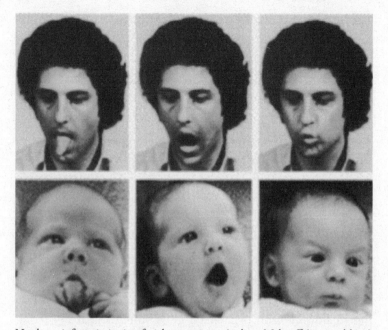

Newborn infants imitating facial expressions. Andrew Meltzoff (pictured here) and his colleagues argue that mirror neurons may help explain how newborns as young as forty minutes old can imitate facial expressions and manual gestures.

monkeys and humans. Many scientists now believe we have neural networks that activate when we perform an action or experience an emotion, and also when we observe someone else performing an action or experiencing an emotion. This might explain why mental states are contagious. It might reveal, at a basic brain level, what happened to me when I saw that NFL ad. Just seeing the goofy smiles on the faces of the football player and the boy triggered an automatic mirror response in my own brain. I literally caught their joy.

Mirror neurons may also be the basis of our ability to run powerful fictional simulations in our heads. A pioneer of mirror neuron research, Marco Iacoboni, writes that movies feel so authentic to us

> because mirror neurons in our brains re-create for us the distress we see on the screen. We have empathy for the fictional characters—we know how they're feeling—because we literally experience the same feelings ourselves. And when we watch the movie stars kiss on screen? Some of the cells firing in our brain are the same ones that fire when we kiss our lovers. "Vicarious" is not a strong enough word to describe the effect of these mirror neurons.

As with any area of emerging science, controversies rage. Some neuroscientists are confident that we understand what is going on in another person's mind by simulating it at a neuronal level—by mirroring that person's brain state with our own. Other scientists are more wary. But whether mirror neurons turn out to be the ultimate explanation or not, we know from laboratory studies that stories affect us physically, not just mentally. When the protagonist is in a tight corner, our hearts race, we breathe faster, and we sweat more. When watching a scary movie, we cringe defensively when the victim is attacked. When the hero brawls with the villain, we may do a little dance in our seats, as though slipping punches. We writhe and weep at *Sophie's Choice*. We laugh until it hurts at *Candide* or *Fear and Loathing in Las Vegas*. We gulp and sweat during the shower scene in *Psycho,* watching through the cracks between our fingers—and the trauma we endure is so real that we may stick to baths for months to come.

In their book *The Media Equation,* the computer scientists Byron Reeves and Clifford Nass show that people respond to the stuff of fiction and computer games much as they respond to real events. For Reeves and Nass, "media equals real life." Knowing that fiction is fiction doesn't stop the emotional brain from processing it as real. That is why we have such a powerfully stupid urge to scream at the heroine of the slasher film, "*Drop the phone and run! Run for God's sake! Run!*" We respond so naturally to fictional stimuli that when psychologists want to study an emotion such as sadness, they often subject people to clips from tearjerkers like *Old Yeller* or *Love Story.*

Our responses to fiction are now being studied at a neuronal level. When we see something scary or sexy or dangerous in a film, our brains light up as though that thing were happening to us, not just to a cinematic figment. For example, in a Dartmouth brain lab, people watched scenes from the Clint Eastwood Spaghetti Western *The Good, the Bad and the Ugly* while their brains were scanned by a functional MRI (fMRI) machine. The scientists, led by Anne Krendl, discovered that viewers' brains "caught" whatever emotions were being enacted on the screen. When Eastwood was angry, the viewers' brains looked angry, too. When the scene was sad, the viewers' brains also looked sad.

In a similar study, a team of neuroscientists led by Mbemba Jabbi placed research subjects in an fMRI scanner while they screened a short clip of an actor drinking from a cup and then grimacing in disgust. They also scanned subjects while the researchers read a short scenario aloud, asking the subjects to imagine walking down the street, accidentally bumping into a retching drunk, and catching some of the vomit in their own mouths. Finally, the scientists scanned the subjects' brains while they actually tasted disgusting solutions. In all three cases, the same brain region lit up (the anterior insula—the seat of disgust). As one of the neuroscientists put it, "What

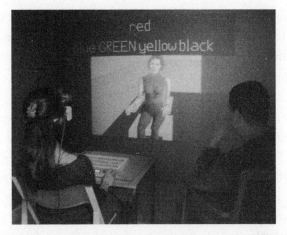

Scientists have begun experimenting on virtual humans in the lab. Here researchers at University College London replicate Stanley Milgram's famous experiments with electric shock—except that they are torturing a semirealistic cartoon instead of a real person. Despite the fact that all of the participants know that the setup is fake, they tend to respond—behaviorally and physiologically—as though the torture is real.

this means is that whether we see a movie or read a story, the same thing happens: we activate our bodily representations of what it feels like to be disgusted—and that is why reading a book and viewing a movie can both make us feel as if we literally feel what the protagonist is going through."

These studies of "brains on fiction" are consistent with the problem simulation theory of storytelling. They suggest that when we experience fiction, our neurons are firing much as they would if we were actually faced with Sophie's choice or if we were taking a relaxing shower and a killer suddenly tore down the curtain.

It seems plausible that our continuous immersion in fictional problem solving would improve our ability to deal with real problems. If this is so, fiction would do so by literally rewiring our brains. It is an axiom of neuroscience that "cells that fire together wire together." When we practice a skill, we improve because repetition of the task establishes denser and more efficient neural connections. This is why we practice: to lay down grooves in our brains, making our actions crisper, faster, surer.

At this point, I should make a distinction between the problem simulator model I'm describing and a related model championed by Steven Pinker. In his groundbreaking book *How the Mind Works,* Pinker argues that stories equip us with a mental file of dilemmas we might one day face, along with workable solutions. In the way that serious chess players memorize optimal responses to a wide variety of attacks and defenses, we equip ourselves for real life by absorbing fictional game plans.

But this model has flaws. As some critics have pointed out, fiction can make a terrible guide for real life. What if you actually tried to apply fictional solutions to your problems? You might end up running around like the comically insane Don Quixote or the tragically deluded Emma Bovary—both of whom go astray because they confuse literary fantasy with reality.

But there's another problem with Pinker's idea. It seems to depend on explicit memory—the type of memory we can consciously access. Yet think back over the past several years. Which novel or film affected you most deeply? Now what do you actually remember about it? If you are like me, you'll re-call a few key characters and the basic gist; sadly, almost all of the granular detail will be lost in an amnesiac fog. And that's for the story that moved you the most. Now think of the thousands of more pedestrian sitcoms, films, and prose fictions you enjoyed over the same period. Almost none of the details will be left in your memory bank.

"A world of disorderly Notions, pick'd out of his books, crouded into his Imagination" (Miguel de Cervantes, *Don Quixote*, 1605–1615).

But the simulator model I'm describing doesn't depend on our ability to store fictional scenarios in an accurate and accessible way. The simulator model depends on *implicit* memory—what our brains know but "we" don't. Implicit memory is inaccessible to the conscious mind. It is behind all the unconscious processing that goes into driving a car, or swinging a golf club, or even navigating the rocky shoals of a cocktail party. The simulator theory is based on research showing that "realistic rehearsal of any skill . . . leads to enhanced performance regardless of whether the training episodes are explicitly remembered." When we experience fiction, our minds are firing and wiring, honing the neural pathways that regulate our responses to real-life experiences.

Researchers have been circling this idea for years, but they've made little progress in testing it. This is not because the idea is untestable. It is because researchers are simply not in the habit of pursuing scientific responses to literary questions. Navy administrators test whether flight simulators work by seeing whether pilots who train on them are more likely to be proficient, not to mention alive, at the end of their training than pilots who flew before the advent of simulators. The evidence is unequivocal: flight simulators work. In the same way, if the evolutionary function of fiction is—at least in part—to simulate the big dilemmas of life, people who consume a lot of fiction should be more capable social operators than people who don't.

The only way to find out is to do the science, and the psychologists Keith Oatley, Raymond Mar, and their colleagues have made a start. In one study, they found that heavy fiction readers had better social skills—as measured by tests of social and empathic ability—than those who mainly read nonfiction. This was not, they discovered, because people who already had good social abilities naturally gravitated to fiction. In a second test that accounted for differences in personality traits—as well as factors such as gender, age, and IQ—the psychologists still found that people who consumed a lot of fiction outperformed heavy nonfiction readers on tests of social ability. In other words, as Oatley puts it, differences in social abilities "were best explained by the kind of reading people mostly did."

Note that these findings are not self-evident. If anything, stereotypes of nerdy bookworms and introverted couch potatoes might lead us to expect that fiction degrades social abilities rather than improving them.

• • •

So here is the central idea as we've developed it to this point. Fiction is a powerful and ancient virtual reality technology that simulates the big dilemmas of human life. When we pick up a book or turn on the TV—whoosh!—we are teleported into a parallel universe. We identify so closely with the struggles of the protagonists that we don't just sympathize with them; we strongly empathize with them. We *feel* their happiness and desire and fear; our brains rev up as though what is happening to them is actually happening to us.

The constant firing of our neurons in response to fictional stimuli strengthens and refines the neural pathways that lead to skillful navigation of life's problems. From this point of view, we are attracted to fiction not because of an evolutionary glitch, but because fiction is, on the whole, good for us. This is because human life, especially social life, is intensely complicated and the stakes are high. Fiction allows our brains to practice reacting to the kinds of challenges that are, and always were, most crucial to our success as a species. And as we'll see, we don't stop simulating when the sun goes down.

Night Story

The pig dreams of acorns, the goose of maize.

— HUNGARIAN PROVERB, quoted in Sigmund
Freud, *The Interpretation of Dreams*

I AM IN MY BED—sweating, panting, paralyzed except
for my eyeballs rolling beneath their lids.

I am in the desert in the last of the dusk. I am walking
hand in hand with my daughter, who is three. The horizon is
far away. The hardpan beneath us is brittle with thirst.

I am sitting high on the rim of a desert canyon, with my
feet dangling down, watching some sporting event way down
in the valley. It is deep night. The athletes are lit with fire that
washes along the canyon walls like waves. I smile down on the
performance, enjoying myself, swinging my legs over the edge
like a boy.

My daughter is in her own world, dancing along the cliff's
edge, twirling and making her skirt and her ponytail fly out.
I watch her singing quietly to herself, twinkling on her toes,

with no need of her heels. I smile at her, noticing and admiring everything, and I turn back to the game.

And then she is falling, and I know I should dive after her, but I can't move.

Despair vaster than I have ever known. Pain beyond enduring or explaining. Death—remorseless, immovable—has her and will never give her back.

But when the dream released me, death released her. For a time, I lay fearfully in the darkness doubting the miracle. Had my most ardent and impossible wish really come true? I lay in bed sending atheist prayers of thanksgiving into the void.

Every night of our sleeping lives, we wander through an alternate dimension of reality. In our dreams, we feel intense fear, sorrow, joy, rage, and lust. We commit atrocities; we suffer tragedies; sometimes we orgasm; sometimes we fly; sometimes we die. While the body lies dormant, the restless brain improvises original drama in the theaters of our minds.

The novelist John Gardner compares fictional stories to

"vivid and continuous dream[s]," but it's just as accurate to call dreams "vivid and continuous stories." Researchers conventionally define dreams as intense "sensorimotor hallucinations with a narrative structure." Dreams are, in effect, night stories: they focus on a protagonist—usually the dreamer—who struggles to achieve desires. Researchers can't even talk about dreams without dragging in the basic vocabulary of English 101: plot, theme, character, scene, setting, point of view, perspective.

Night story is a mystery. Who hasn't marveled at the psychotic creativity of their own dreaming mind? What does it mean when our dreaming selves appear naked in church, or murder some innocent, or soar through the sky? What does it mean when a dream finds you in a bathroom, staring in stark terror as an evil elf masturbates above the clothes hamper, and your mother is pounding on the door, and you look in the mirror and there is no elf crouching over the clothes hamper, only you? Who has not wondered, as I did in the sweaty aftermath of my desert dream, why his brain decided to stay up all night just to torture itself? Why do we dream?

For thousands of years, dreams were explained as encrypted messages from the spirit world that could be decoded only by priests and shamans. And then, in the twentieth century, Freudians triumphantly announced that dreams were actually encrypted messages from the id, and only the priesthood of psychoanalysts could decode them. To get the flavor of psychoanalytic dream interpretation, consider Frederick Crews's gloss of one of Freud's most famous case studies, that of the Wolf Man, Sergei Pankeev (or Pankejeff).

> Freud was determined to find a primal scene to serve as the fountainhead of Pankeev's symptoms. He made it materialize through a transparently arbitrary interpretation of a remembered dream of Pankeev's from the suspiciously early age of four, about six or seven white wolves (actually dogs, as Freud was later compelled to admit) sitting in a tree outside his window. The wolves, Freud explained, were the parents; their whiteness meant bedclothes; their stillness meant the opposite, coital motion; their big tails signified, by the same indulgent logic, castration; daylight meant night; and all this could be traced most assuredly to a memory from age one of Pankeev's mother and father copulating, doggy style, no fewer than three times in succession while he watched from the crib and soiled himself in horrified protest.

Nowadays many dream scientists look with bald disdain on the Freudian legacy. The most prominent dream scientist of the modern era, J. Allan Hobson, calls psychoanaly-

Sigmund Freud (1856–1939). This photograph was taken around the time he published his most famous work, *The Interpretation of Dreams* (1900).

sis a "fortune cookie" model of dream interpretation. For researchers like Hobson, the search for a hidden symbolology of dreams has been a tragic waste of time. Taking up Freud's contrast between the manifest (obvious) and the latent (symbolic) content of dreams, Hobson shouts, "The manifest dream is the dream, is the dream, is the dream!" But what are dreams for, if they are not coded messages from the gods or from one's own psyche?

Dreams, like hands, may have multiple functions. For example, there is some evidence that dreams may help us file new experiences in the correct bins of short- and long-term memory. Many psychologists and psychiatrists believe that dreams may also be a form of autotherapy, helping us to cope with the anxieties of our waking lives. Or, as the late Nobel laureate Francis Crick proposed, dreams may help us weed useless information out of the mind. For Crick, dreams were a disposal system: "We dream to forget."

Still others believe that dreams have no purpose whatsoever. As the dream researcher Owen Flanagan puts it, "Our dreams were not designed by nature to serve any function . . . Nothing, nada, just noise, like the gurgling of the stomach or the sound of the heart." The thudding sound of the heart is not the point of pumping the blood. As the children's book says, everyone poops, but this is not the point of eating. Defecation is a side effect of our need to eat. In the same way—according to Flanagan and many other researchers—a dream is just brain waste. It is a useless by-product of all the useful work the sleeping brain does. Why did I dream my desert dream? For no reason at all.

The by-product theory of dreams goes by the acronym RAT (random activation theory). RAT is based on the idea that the brain has serious work to do at night, especially during REM sleep. This night work may be one of the reasons we sleep in the first place: so the brain can finish all the housekeeping chores it can't get to during the day.

But as we will see in the next chapter, we all have brain circuits that pore over incoming information, filter for patterns, and arrange those patterns into narratives. According to RAT, these storytelling circuits have a flaw: they never learn that the clatter and noise of the sleeping brain is meaningless. Instead, they treat that clatter just like the information that streams in during waking hours, trying to turn it into a coherent narrative. Our inner storyteller does this for no practical reason. It just does it because it has a lifelong case of insomnia and because it is addicted to story—it simply can't help it.

RAT is a bold theory that, in defining dreams as mental garbage, challenges all the canons of psychology and folk wisdom. And RAT is also a clever theory that organizes a lot of chaotic data about dreams. Why are dreams so strange? Why do fathers sit placidly as their precious daughters pirouette on a cliff's edge? Why do elves masturbate angrily in your bathroom? All of the bizarre elements of dreams can be chalked up to the mind's desperate attempts to craft ordered narratives out of random input.

While everyone agrees that some features of dreams may be by-products, support for RAT is far from unanimous. RAT critics argue that although dreams are certainly strange, they are not strange enough to be explained by RAT. The Finnish dream researcher Antti Revonsuo believes we are too easily impressed by the weirdness of dreams. We remember our bizarre dreams best, and so we fail to register that dreams are mainly realistic and coherent. Of RAT theory, Revonsuo writes:

> No random process could ever create such a complex simulation of the waking world. Bizarreness certainly is a regular fea-

ture of dreams, but it constitutes a relatively mild deviation against the solid background of sophisticated organization; a little noise within a highly organized signal. RAT can only explain the bizarreness — the small degree of noise — but it cannot explain the high degree of organization in which the bizarreness is embedded.

Revonsuo also points out that RAT's proponents have an unwarranted confidence in the arrow of causation. RAT simply assumes that certain brain states cause the properties of dreams. For instance, RAT advocates say that dreams are highly emotional because the limbic system and the amygdala — both of which are linked to emotion — happen to be aroused during REM sleep. RAT doesn't allow for an equally plausible interpretation: the emotional centers of the brain are aroused *because* the dreamer is dreaming emotionally.

And then you have atonia, the sleep paralysis that sets in during REM sleep. Why do we have it? It must be because, eons ago, our ancestors were harming themselves and others by acting out their dreams. When you have a dream brawl with a flesh-eating zombie, your brain doesn't know it is dreaming. It thinks it is actually fighting the zombie. It is flooding the body with commands: *Crouch! Left jab! Right cross! Poke it in the eye! Run for it! Run!* The only reason our sleeping bodies do not obey these commands is that they never receive them. All the motor commands are being intercepted by a blockade in the brain stem.

But the blockade doesn't make good sense from a RAT perspective. In RAT, certain parts of our brains concoct crazy fictions out of the sleepy blabber generated by other parts of our brains. No problem, except that the motor centers of the brain have an alarming tendency to treat the night story as absolutely real and to respond appropriately—to leap when the dream says leap, to fight and flee when the dream calls for it. So, from the RAT point of view, dreams are actually much worse than merely worthless; they are dangerous.

The simplest evolutionary solution to this problem would not be the drastic one of paralyzing the whole body during REM dreams. The simplest solution would be to silence our inner storyteller—to disable it for the night, or to temporarily block its access to the rest of the brain's doings. Instead, evolution designed a solution that allows the mind to safely run its simulations, as though dreams serve an important purpose that needs to be protected.

But perhaps most damaging to RAT is the evidence that dreaming is not limited to humans but is widespread in other animals. REM, dreams, and the brain blockades that make dreaming safe have apparently been conserved across widely differing species, suggesting that dreams have value.

JOUVET'S CATS

Back in the 1950s, the French dream scientist Michael Jouvet knew that many animals experience REM sleep and atonia. But did they actually dream? In hopes of finding out, Jouvet procured a large number of stray cats, along with bone saws, slicers, and stabbers. He cut open the heads of his cats, located the brain stems, and began screwing up the works.

Jouvet was trying to destroy the blockade-making capacity of kitty atonia. If cats dreamed, he reasoned, motor signals from the brain would reach the muscles, and the cats would act out their dreams. By trial and error, Jouvet had to do enough damage to the feline brain stem to destroy the machinery of atonia without otherwise impairing his cats or killing them.

Jouvet placed his surviving cats in big Plexiglas boxes. He hooked them up to devices that monitored their vital signs and brain activity. A camera recorded them while they slept. Jouvet's results were remarkable and unequivocal: kitties dream. Some minutes after entering REM sleep, Jouvet's cats would appear to suddenly wake up. Their eyes would flutter open, and they would raise their heads and look around. Their eyes were open, but they couldn't see. They did not respond to offers of delicious food. When the researchers flicked their hands toward the cats' eyes, the cats did not flinch.

The cats were far away in Dreamland, totally oblivious to the waking world. They were acting out their dreams. They moved around and reconnoitered. They took up the postures of stalking or lying in wait. They engaged in predatory behavior, springing on invisible prey and plunging their teeth into their victims. And the cats also exhibited defensive behavior: they would flee from threats by backpedaling with flattened ears, or they would fight threats with hisses and paw swipes.

In short, Jouvet's experiment showed not only that cats dream but also that they dream about very specific things. He pointed out the obvious: a cat "dreams of actions characteristic of its own species (lying in wait, attack, rage, fright, pursuit)." But look at Jouvet's list. In fact, he didn't find that cats dream about the "characteristic" actions of their own species. Instead, his cats seemed to dream about a narrower subset of *problems* in kitty life—namely, how to eat and not be eaten.

For your average tomcat, Dreamland is not a world of catnip debauches, warm sunbeams, canned tuna fish, and yowling bitches in heat. Dreamland for cats is closer to kitty hell

Scientists at MIT determined that rats probably dream, which is damaging to RAT.

than kitty heaven, as feelings of fear and aggression predominate.

Jouvet thought his research had implications beyond the feline. He considered it unlikely that dreams were rare in animals and that he had just happened to find them in the one species where he looked. Jouvet believed that dreams were probably common across the animal kingdom and that they serve a purpose.

Jouvet raised the possibility that dreams are for practice. In dreams, animals rehearse their responses to the sorts of problems that are most germane to their survival. Kitties practice on kitty problems. Rats practice on rat problems. Humans practice on human problems. The dream is a virtual reality simulator where people and other animals hone responses to life's big challenges.

As a general theory of dreams, Jouvet's idea did not immediately catch on, partly because, in the 1950s, the whole field of dream research was still overshadowed by Freud. For Freudians, dreams are disguised fulfillments of repressed wishes—thus Freud's approving quotation of the proverb "The pig dreams of acorns, the goose of maize." But Jouvet's theory, supported by a number of studies since, is close to the opposite of Freud's wish fulfillment theory. Recent research suggests that if geese dream—and it is possible that they do—they probably don't dream of maize. They probably dream of foxes. And cats dream of snarling dogs. And people dream of bad men, and of little girls losing their balance and falling through space.

PEOPLE DREAM OF MONSTERS

Dreamland is a lot different than most of us think. Freud's theory of dreams as disguised wishes just formalizes the popular idea that Dreamland is, well, dreamy. "Did you enjoy yourself?" you ask your colleague upon her return from vacation, "Oh, yes, it was a dream!" A wonderful thing happens, and you call it "a dream come true."

But thank goodness dreams usually don't come true. As J. Allan Hobson put it, "[In dreams] waves of strong emotion—notably fear and anger—urge us to run away or do battle with imaginary predators. Fight or flight is the rule in dreaming consciousness, and it goes on and on, night after night, with all too rare respites in the glorious lull of fictive elation." Although there is some controversy about how to interpret the data, most dream researchers generally agree with Hobson: Dreamland is not a happy place.

Dream reports are collected in two main ways. In the less controlled method, subjects keep a dream diary. When they wake up in the morning, they immediately roll over and record their dreams before they fade. In the more controlled method, sleeping subjects are jostled awake in the lab, and the researcher says, "Quick, what were you dreaming about?" Patterns in dream reports are then quantified using various techniques of analysis.

Consider my dreams from the night of December 13, 2009, which I rushed to record upon waking:

- I dreamed it was Christmas morning. I woke up and realized that I had neglected to buy my wife a present. As we began to open presents everyone else in the family was happy, but I was perishing from anxiety: how do I get out of this one?
- Another anxiety dream: a writer scooped me on an important portion of this book, beating me to press and rendering my work useless.
- I was out jogging, pushing my sleeping daughter in her stroller. Annabel awoke, groggily rolled out of the stroller, and slid down a paved embankment. Terrified, I rushed down after her. Her back was scraped up, but she was otherwise fine. A rich woman in a fancy car pulled off the road and gave us a ride back to her huge mansion, where Annabel romped with other children in a spectacular indoor playground.
- I was at a resort, possibly the one where my wife and I honeymooned. An attractive woman was trying to have an affair with me. Before I could decide what to do, I learned that it was all a setup to blackmail me.

None of the above dreams was quite a nightmare. None of them marked me in the way that my desert dream did. But they were intensely anxious dreams where the most important things in my life were threatened: the love of my wife; my work as a writer; the safety of my child; my reputation for basic decency (if not virtue).

Although we obviously can't perform Michael Jouvet's experiments on people, nature has in fact performed the experiment many times. REM behavior disorder (RBD) mainly afflicts elderly men with certain neurodegenerative disorders, such as Parkinson's disease. The disease riddles the brain and prevents dream blockades from forming in the brain stem. While their brains sleep and dream, men with RBD rise from bed to act out their brains' commands. As with Jouvet's cats, the men seem very rarely to dream of happy things. They dream instead of trouble. In one study of four men with RBD, all four manifested dangerously aggressive behavior in their sleep, sometimes harming themselves or their wives.

My miniature dream diary is consistent with research studies showing that Dreamland is rife with emotional and physical peril. In a 2009 review of the threatening aspects of dreams, Katja Valli and Antti Revonsuo lay out some astonishing statistics. Valli and Revonsuo estimate that an average person has 3 REM dreams per night and about 1,200 REM dreams per year. Based on analysis of dream reports across many large studies, they estimate that 860 of those 1,200 dreams feature at least one threatening event. (The researchers define "threat" broadly, but reasonably, as a physical threat, a social threat, or a threat to valuable possessions.) But since most threatening dreams portray more than one threatening event, the researchers estimate that people experience about 1,700 threatening REM dream episodes per year, or almost 5 per night. Projected over a seventy-year life span, the average person would suffer through about 60,000 threatening REM dreams featuring almost 120,000 distinct threats. In short, what Vivian Paley wrote of children's pretend dramas seems to apply just as well to dreams: they are "the stage on which bad things are auditioned."

Valli and Revonsuo acknowledge that their statistics are far from ironclad. They are projections based on imperfect data. But whether these threat estimates are off by a little or a lot, they still establish an important point: Dreamland is, incontestably, far more threatening than the average person's waking world. As Revonsuo writes, threats to life and limb "are so rare in real life that if they occur with almost any frequency at all in dreams, they are very likely to exceed the frequency of occurrence in real life." For example, the Finnish college students who serve as subjects in Valli and Revonsuo's research do not confront bodily threats on a *daily* basis, but they do confront them on a *nightly* basis.

It is important to stress that the same threat patterns have emerged not only in Western college students but in all populations that have been studied—Asians, Middle Easterners, isolated hunter-gatherer tribes, children, and adults. Around the world, the most common dream type is being chased or attacked. Other universal themes include falling from a great height, drowning, being lost or trapped, being naked in public, getting injured, getting sick or dying, and being caught in a natural or manmade disaster.

It's therefore unsurprising that the dominant emotions

in Dreamland are negative. When you are visiting Dreamland, you may sometimes feel happy, even elated, but mostly you feel dragged down by anger, fear, and sadness. While we sometimes dream of thrilling things, such as sex or flying like a bird, those happy dreams are much rarer than we think. People fly in only one out of every two hundred dreams, and erotic content of any kind occurs in only one in ten dreams. And even in dreams where sex is a major theme, it is rarely presented as a hedonistic throw down. Rather, like our other dreams, sex dreams are usually edged with anxiety, doubt, and regret.

THE RED THREAD

Just as conflict and crisis are greatly overrepresented in dreams, the ordinary stuff of life is underplayed. For example, researchers studied dream reports from four hundred people who spent an average of six hours per day absorbed in the minutiae of office and student life—typing, reading, calculating, working on a computer. And yet the workaday activities that dominated their waking hours almost never featured in their dreams. Instead, they dreamed of trouble. Trouble is the fat red thread that ties together the fantasies of pretend play, fiction, and dreams, and trouble provides a possible clue to a function they all share: giving us practice in dealing with the big dilemmas of human life.

A conservative estimate, accounting only for REM dreams, would suggest that we dream in a vivid and storylike way for about two hours per night, which comes to fifty-one thousand hours over an ordinary life span, or about six solid years of nonstop dreaming. During these years, our brains simulate many thousands of different responses to many thousands of different threats, problems, and crises. Crucially, there is usually no way for our brains to know that the dream is just a dream. As sleep researcher William Dement put it, "We experience a dream as real because," from the brain's perspective, "it *is* real."

The University of Michigan psychologists Michael Franklin and Michael Zyphur think these facts have important implications:

> When you consider the plasticity of the brain—with as little as 10–12 minutes of motor practice a day on a specific task [say, piano playing] the motor cortex reshapes itself in a matter of a few weeks—the time spent in our dreams would surely shape how our brains develop, and influence our future behavioral predispositions. The experiences we accrue from dreaming across our life span are sure to influence how we interact with the world and are bound to influence our overall fitness, not only as individuals but as a species.

But what about our poor memories? Franklin and Zyphur acknowledge that dream amnesia is a highly intuitive reason for dismissing dreams. Since memories of our dreams are usually burned away with the morning light, they can't be worth much to us.

But as we've already seen, conscious knowledge can be overrated. There are two kinds of memory: implicit and explicit. The problem simulation model is based on implicit, unconscious memory. We learn when the brain rewires itself, and we don't need to consciously remember for that rewiring to occur. The most spectacular proof of this comes from amnesia victims, who can improve by practicing tasks without retaining any conscious memory of the practice.

I recently "taught" my older daughter, Abby (six at the

time), how to ride her bike without training wheels. I say "taught" because all I actually did was jog, dadlike, at her side as she wobbled down the road, teaching herself how to balance. Within a week or so, Abby pretty much had it down, and she was turning tight circles in the road. I was impressed that she had learned how to turn, and I asked her how she had done it. She answered confidently, "I just twist the handles, and it turns the wheel."

A reasonable response, but that's not how Abby turns her bike. As the Berkeley physicist Joel Fajans explains, turning a bike is actually a complex procedure: "If you attempt to make a right turn on your bicycle before first leaning your bicycle over to the right, centrifugal forces will cause you to crash by falling over to the left . . . Leaning the bicycle to the right allows gravity to cancel the centrifugal forces. But how do you get the bike to lean to the right? By countersteering, i.e. by turning the handlebars to the left. In other words to make a right turn, you first turn the handlebars left!"

Someday Abby may have no memory of learning to ride her bike—no memory of her fear or her pride, or of me puffing along at her side. But she'll still know how to ride it. Bike riding is an example of how we can learn something, and learn it well, without our conscious minds having a clue. Our brains know a great deal that "we" don't.

Skeptics such as the psychologist Harry Hunt raise what is, on the face of it, a more powerful objection to the problem simulation theory of dreams. If a simulator is going to be worth anything, it has to be realistic. For instance, a flight simulator that is insufficiently realistic will train aviators, but that training will be a curse, not a benefit. Hunt and others have argued that dreams can't function as simulators because they are unrealistic. "It is difficult to see," Hunt writes, "how our paralyzed fears, slow motion running, and escape tactics based on absurd reasoning could be a rehearsal of anything adaptive."

But look at Hunt's objection. He is not describing dreams generally; he is describing nightmares. Studies show that nightmares are more bizarre than other kinds of dreams and that we remember them better. So the idea that dreams are hopelessly bizarre may be a side effect of the sorts of dreams we are most likely to remember. Indeed, large samples of dream reports from multiple studies suggest that the majority of dreams are reasonably realistic. Valli and Revonsuo conclude that our responses to dream dilemmas are usually "relevant, reasonable, and appropriate to the dreamed situation."

According to Revonsuo, the simulation model of dreams represents a gigantic breakthrough: "We are for the first time in a position to truly understand why we dream." But as I've stressed throughout this book, identifying *a* function for dreams or pretend play or fiction doesn't mean that we've identified *the* function. My desert nightmare may have been a way of priming me to take better care of what is most dear to me. But that's not all it was. I'm ashamed to admit it, but the dream was an exaggerated version of a typical scene: my daughters trying, with only partial success, to get my attention as I read, write, or watch the big game on TV. My nightmare was, in part, a sharp rebuke. The dream was saying, The big game is not important. The book on your lap is not that important. Pay attention: it's the girl who's important.

The Mind Is a Storyteller

Man—let me offer you a definition—is the storytelling animal.
Wherever he goes he wants to leave behind not a chaotic wake,
not an empty space, but the comforting marker buoys and trail
signs of stories. He has to keep on making them up. As long as
there's a story, it's all right. Even in his last moments, it's said, in
the split second of a fatal fall—or when he's about to drown—he
sees, passing rapidly before him, the story of his whole life.

— GRAHAM SWIFT, *Waterland*

I T IS DECEMBER 30, 1796. A gang of criminals is gathered in a big, dank cellar deep below Bethlem Hospital in
London. The gloomy leader of the gang, Bill the King,
makes a circuit of the cellar, inspecting his terrible machine.
He toes the big wooden kegs with their steel hoops. The kegs
are full of nasty stuff such as human seminal fluid, dog fe-
ces, the stench of the cesspool, and horse farts. Bill the King
presses his ear to the kegs and hears the stuff churning and
percolating under pressure. He fingers the hissing tubes that
run like black serpents from the kegs, spraying noxious fumes
into the guts of the machine.

As Bill the King examines the device, he passes the other villains. He sees young Charlotte lying on the cold cobbles, half-naked, raw from her chains. He sees Jack the Schoolmaster penciling figures in his notebook. The Schoolmaster calls out to Bill with an inside joke, "I'm here to see fair play!" Bill the King walks on, scowling; no man or woman has ever yet made Bill smile. But Sir Archy smiles, and laughs, too, calling out to the Schoolmaster with a joke of his own—something about young Charlotte, something vile. Sir Archy has a high voice and always fakes an accent. The others suspect he's a woman in drag.

Bill the King ignores the fools. He circles the great engine, running his hand along the oaken grain, eyeing every board and bolt. The machine is called an air loom. It is a great square box on squat legs. It is six feet tall and fifteen feet per side.

Bill the King reaches the air loom's control panel, where a pockmarked hag is sitting. Sir Archy and Jack the Schoolmaster tease her mercilessly. They call her the Glove Woman. The Glove Woman sits at a stool and works the air loom's controls with blurry speed. Her gloved hands are everywhere at once, twisting and pulling at knobs, flipping switches, and tickling the ivory keys of an organ. Bill the King leans down and whispers into her ear, "Fluid locking."

The Glove Woman twists the biggest brass knob. The pointer rattles past settings labeled as follows: KITEING (which suspends an idea in the mind for hours), LENGTH-ENING OF THE BRAIN (which twists the thoughts), GAZ-PLUCKING (which harvests the magnetic fluids in farts from the anus), BOMB-BURSTING (painful explosions in the head), THOUGHT-MAKING and FOOT-CURVING (which are what they sound like), VITAL-TEARING and FIBRE-RIP-

PING (which are very painful), and, gravest of all, LOBSTER-
CRACKING (which is lethal). The Glove Woman settles the
pointer on FLUID LOCKING, and Bill murmurs, "Proceed."

The Glove Woman knows nothing of the science of the air
loom, but she has spent many years playing the machine, and
now she is a virtuoso. She works the levers and operates her
organ. Each organ key regulates the valve to a different keg.
Each setting of the air loom is like a song composed of electrochemical notes.

Inside the machine, under the vibrating influence of the
magnets, the great loom begins to "weave" the chemical es-
sences from the kegs directly into the air. The magnetized gas
flows upward and fills the sails of a small, windmill-like device
that sits on the air loom's roof. The sails of the device begin to
turn, projecting the unnatural gas outward and upward like a
ray.

The magnetized gas exits through the cellar wall and
rises through many fathoms of wet soil and rubble; it passes
through the floor of the House of Commons and focuses on a
certain young man sitting in the spectators gallery. That young
man, James Tilly Matthews, suddenly feels the air churning
around him. He tastes blood in his mouth. He knows what
to do. He holds his breath for a long time. He randomizes his
breathing pattern and, in so doing, hides from the air loom.

Matthews sits, breathing raggedly, watching the great men
of his age bellowing and preening on the floor. Prime Minister William Pitt stands and rails against France, pounding the
drums of war. Pitt's performance confirms Matthews's worst
fears: the prime minister of England is the Air Loom Gang's
puppet.

Matthews is dismayed. He's the only one who knows the
truth: the government of England has been taken over by a

Artist Rod Dickinson's re-creation of the air loom, faithfully based on James Tilly Matthews's own sketch.

powerful conspiracy. Matthews—a ruined tea merchant, a pauper—is living at the very center of history, and the Air Loom Gang knows it. That is why the Duke of York and the king of Prussia are moving darkly against him. That is why draconian laws have been passed to keep him quiet. That is why whole government agencies exist to intercept his mail.

Matthews makes a grave error. He lets his worries distract him. He forgets to hide from the machine. He breathes deeply and evenly, and the air loom finds him, brimming his lungs with the vile, prickly gas. Matthews knows it is fluid locking. He can feel the gas bubbling in his veins, curdling his fluids, freezing the muscles at the base of his tongue. In seconds, he will lose the power of speech. He must act now. He stands up and screams down at the members of Parliament, "Treason!"

THE CRAZED OF THE CRAFT

James Tilly Matthews was confined as an incurable lunatic in Bethlem Hospital, better known as Bedlam, although he did have several doppelgängers who moved freely outside the hospital walls. While dodging air loom rays at Bedlam, Matthews realized that he was the emperor of the world, and he wrote bitter complaints against the kings and potentates who had usurped him.

Matthews's doctor was named John Haslam, and he described Matthews's case in a book called *Illustrations of Madness*. (The preceding account is drawn mainly from Haslam's book and from Mike Jay's fascinating study of Matthews, *The Air Loom Gang*.) *Illustrations of Madness* is a classic case study in the field of mental health, and the first clear description of paranoid schizophrenia in the annals of medicine.

Schizophrenia has been called "the central mystery of psychiatry." Schizophrenics are prey to a variety of bizarre beliefs, delusions, and hallucinations. Like Matthews, they often hear alien voices muttering in their ears. They often believe that their actions are being orchestrated by outside forces—by extraterrestrials, gods, demons, or wicked conspiracies. And they often have delusions of grandeur: they think they are important enough for aliens, demons, and conspirators to target.

James Tilly Matthews's vision of the Air Loom Gang was a bravura work of fiction. He made himself the struggling protagonist in a sweeping drama with world-historical implications, giving bit roles to real potentates and prime ministers. Matthews also invented a whole cast of fully realized villains. Bill the King, the Glove Woman, and Sir Archy have all the quirks and tics that turn flat characters round. If Matthews had fashioned his conspiratorial delusions into a novel, he might have made a lot of money. He might have been an eighteenth-century Dan Brown.

When Matthews was about thirty years old, his brain decided, without his permission, to create an intricate fiction, and Matthews spent the rest of his life living inside. It is tempting to draw analogies between the creativity of the delusional schizophrenic and that of the creative artist. And indeed, for hundreds of years, a relationship between madness and artistic genius has been a kind of cultural cliché. Lord Byron wrote of poets, "We of the craft are all crazy." John Dryden's poem "Absalom and Achitophel" proclaims, "Great wits are sure to madness near allied, / And thin partitions do their bounds divide." And Shakespeare wrote in *A Midsummer Night's Dream* that lunatics and poets are "of imagination all compact."

For a long time, it was possible to dismiss links between madness and creativity as purely anecdotal: Vincent van Gogh carving off his ear, Sylvia Plath gassing herself in her oven, Graham Greene playing Russian roulette, Virginia Woolf stuffing her pockets full of rocks and taking a last swim in the river Ouse. But over the past several decades, stronger evidence has accumulated.

The psychologist Kay Redfield Jamison, who has written movingly about her own struggles with bipolar disorder, argues for a strong connection between mental illness and literary creativity in her classic book *Touched with Fire*. In studies of deceased writers—based on their letters, medical records, and published biographies—and in studies of talented living writers, mental illness is prevalent. For example, fiction writers are fully ten times more likely to be bipolar than the general population, and poets are an amazing forty times more likely to struggle with the disorder. Based on statistics like these, psychologist Daniel Nettle writes, "It is hard to avoid the conclusion that most of the canon of Western culture was produced by people with a touch of madness." Essayist Brooke Allen does Nettle one better: "The Western literary tradition, it seems, has been dominated by a sorry collection of alcoholics, compulsive gamblers, manic-depressives, sexual predators, and various unfortunate combinations of two, three, or even all of the above."

In psychiatrist Arnold Ludwig's massive study of mental illness and creativity, *The Price of Greatness,* he found an 87 percent rate of psychiatric disorders in eminent poets and a 77 percent rate in eminent fiction writers—far higher than the rates he found among high achievers in nonartistic fields such as business, science, politics, and the military. Even college students who sign up for poetry-writing seminars have more bipolar traits than college students generally. Creative writers are also at increased risk of unipolar depression and are more likely to suffer from psychoses such as schizophrenia. It is, therefore, not surprising that eminent writers are also much more likely to abuse alcohol and drugs, spend time in psychiatric hospitals, and kill themselves.

Could it be that something about the writer's life—the loneliness, the frustration, the long rambles through imagination—actually triggers mental illness? Possibly. But studies of the relatives of creative writers reveal an underlying genetic component. People who are mentally ill tend to have more artists in their families (especially among first-degree relatives). And artists tend to have more mental illness in theirs (along with higher rates of suicide, institutionalization, and drug addiction).

In the wild fantasies of James Tilly Matthews, we see a dis-

In his memoir, Stephen King writes that he is skeptical of the "myth" associating substance abuse and literary creativity. Yet before getting sober, King drank a case of beer a day and wrote *The Tommyknockers* with cotton swabs stuffed up his nose to "stem the coke-induced bleeding." At his intervention, King's wife dumped his office trash can on the floor. The contents included "beercans, cigarette butts, cocaine in gram bottles and cocaine in plastic Baggies, coke spoons caked with snot and blood, Valium, Xanax, bottles of Robitussin cough syrup and NyQuil cold medicine, even bottles of mouthwash."

eased mind franticly trying to weave sense into the gobbledygook it is inventing—a story that will give structure to the visions, the alien voices, the convictions of personal grandeur. We may shake our heads at Matthews's bizarre delusions, but we are all more like him than we know. Our minds, too, constantly struggle to extract meaning from the data rivering through our senses. Although the stories that sound-minded people tell themselves rarely go awry in the spectacular fashion of paranoid schizophrenics', they often do go awry. This is part of the price we pay for having storytelling minds.

SPLITTING THE BRAIN

In 1962, the storytelling mind was inadvertently isolated—if not discovered—when a neurosurgeon named Joseph Bogen persuaded a severely epileptic patient to undergo a dangerous experimental procedure. Bogen drilled small holes through the top of the patient's skull, introduced a specialized saw into one of the holes, and used it to open a "bone flap" in the cranium. Using scissors, Bogen cut through the brain's protective leathery covering, the dura. Then he gingerly pried apart the two lobes of the patient's brain until he could see the corpus callosum, the nervy band of fibers that routes information back and forth between the left and right hemispheres. Like a saboteur knocking out communication lines, Bogen introduced a scalpel and severed the corpus callosum. In effect, he split the brain—the left and right hemispheres could no longer communicate. Then he reconnected the bone flap with tiny screws, stitched up the patient's scalp, and waited to see what would happen.

Bogen was no mad scientist. The operation was dangerous and its outcome uncertain. But Bogen's patient, a former paratrooper, was dealing with life-threatening seizures that were unresponsive to standard treatments. Bogen believed that severing the corpus callosum might control the seizures. Animal studies suggested it might, and Bogen knew of human cases where seizures had abated after tumors or injuries harmed the corpus callosum.

Perhaps surprisingly, Bogen's surgical Hail Mary pass was

a success. Although his epileptic patient still experienced sei-zures, their frequency and severity were greatly diminished. And even more surprisingly, there seemed to be no side effects. The split-brained man reported no differences in any of his mental processes.

In the days before fMRI machines and other advanced methods of brain imaging, split-brain patients were a boon to neuroscience. Thanks largely to these patients, scientists were able to isolate and study the workings of the two hemispheres of the brain. They discovered that the left brain is specialized for tasks such as speaking, thinking, and generating hypotheses. The right brain is incapable of speech or serious cognitive work; its jobs include recognizing faces, focusing attention, and controlling visual-motor tasks.

The leading pioneer of split-brain neuroscience is Michael Gazzaniga. In his research, Gazzaniga and his collaborators have identified specialized circuitry in the left hemisphere that is responsible for making sense of the torrent of information that the brain is always receiving from the environment. The job of this set of neural circuits is to detect order and meaning in that flow, and to organize it into a coherent account of a person's experience—into a story, in other words. Gazzaniga named this brain structure "the interpreter."

Because of the quirky wiring of the brain, visual information that enters the right eye is fed to the left brain, and information entering the left eye goes to the right brain. In an intact brain, visual information that goes to the left brain is then piped via the corpus callosum to the right brain. But in split-brain individuals, information that enters only one eye gets marooned in the opposite hemisphere, leaving the other hemisphere in the dark.

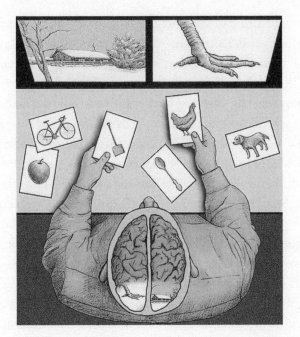

In a brilliant series of experiments, Gazzaniga and his colleagues had split-brain subjects stare at a dot in the center of a computer screen. They then flashed images to the right and left of the dot. Images flashed to the left of the dot were piped only to the right brain, while images that appeared to the right of the dot were sent only to the left brain.

In one experiment, Gazzaniga and his colleagues showed a chicken claw to a split-brain subject's left brain and a snowy scene to his right brain. They then asked the subject to select from an array of pictures lined up in front of him. Again, due to the odd way the brain is wired, the right side of the human body is predominantly controlled by the left brain and the left side by the right brain. With the right hand, the subject chose a picture of a chicken (because the side of the brain that controls that hand had seen a chicken claw). With the left hand, the subject chose a picture of a snow shovel (because the side of the brain controlling that hand had seen a snowy scene).

The split-brain subject was then asked why he chose those two images. The first part of the subject's response made perfect sense: "I chose the chicken, because you showed me a picture of a chicken foot." The subject was able to respond correctly because the image of the chicken claw had been fed to the left hemisphere, which is the verbal side of the brain. But the right side of the brain is mute. So when the subject was asked, "Why did you choose the shovel?" he was not able to give the correct response: "Because you showed me a picture of a snowy scene."

All of this zigzagging from right to left might be a little confusing. But bear with me—the underlying point is simple. The side of the subject's brain responsible for communicating with the researchers (the left) had no idea that the right side of the brain had received an image of a snowy scene. All the speaking side of the brain knew was that the left hand (controlled by the right brain) had reached out and chosen a picture of a snowy scene. It had no idea why. Nonetheless, when the researchers asked, "Why did you choose the shovel?" the subject had a ready and confident response: "Because you need a shovel to clean out the chicken coop."

Gazzaniga and his colleagues varied these studies in all sorts of clever ways. When they fed a split-brain subject's right hemisphere a funny image, the subject would laugh. A researcher would then ask, "Why are you laughing?" The subject's left brain, which was responsible for answering the question, had absolutely no idea. It was not in on the joke. But that didn't stop the left brain from inventing an explanation. The subject might claim that he had just remembered a funny incident. In another study, a subject's right brain was flashed an image of the word "walk." The subject stood up obediently and started walking across the room. When a researcher asked where the man was going, he spontaneously fabricated—and believed—a story about being thirsty and wanting a Coke.

These are representative examples of a pattern Gazzaniga and his colleagues exposed again and again in split-brain subjects. The left brain is a classic know-it-all; when it doesn't know the answer to a question, it can't bear to admit it. The left brain is a relentless explainer, and it would rather fabricate a story than leave something unexplained. Even in split-brain subjects, who are working with one-half of their brains tied behind their backs, these fabrications are so cunning that they are hard to detect except under laboratory conditions.

If Gazzaniga's research just applied to the few dozen epileptics who have undergone split-brain surgery, it would be of limited interest. But this research has important implications for how we understand ordinary, intact brains. The storytelling mind is not created when the scalpel cuts into the corpus callosum. Splitting the brain just pins it down for study.

SHERLOCK HOLMES SYNDROME

You can think of your own storytelling mind as a homunculus (a tiny man) who dwells perhaps an inch or two above and behind your left eye. The little man has a lot in common with Sherlock Holmes, the great literary patriarch who paved the way for a thousand fictional detectives, including the forensic whizzes of hit television shows such as *CSI*. In Sir Arthur Conan Doyle's portrait, Holmes is a genius of criminal investigation, a Newton of the new science of criminology. Holmes has a spooky ability to look at a certain outcome—a corpse,

Illustration from the 1906 edition of Sir Arthur Conan Doyle's *A Study in Scarlet.*

a smattering of clues—and see the whole rich story that led up to it: a love affair, poison pills, adventures in the American West with Brigham Young and the Mormons.

These details come from the first Sherlock Holmes novel, *A Study in Scarlet* (1887). The novel begins by introducing the narrator, ("my dear") Watson—who is not so much a character as a literary device—whose job it is to highlight Holmes's brilliance through his own conventionality. Watson first meets Holmes in a smoky chemistry lab, where the genius is perfecting new forensic techniques. Holmes—tall, lithe, haughty—turns to Watson and shakes his hand. And then, for the first of a thousand times, the wizard blows Watson's mind. He says, "You have been in Afghanistan, I perceive."

Watson is dumbstruck. How could Holmes have known? Later, when Holmes and Watson are lounging in their bachelor pad, Holmes explains that there was no magic in his insight, only logic. With great relish, he tells Watson how he "reasoned backwards" from the silent details of his appearance to make rational inferences about Watson's life. "The train of reasoning," Holmes says, ran like this:

> Here is a gentleman of a medical type, but with the air of a military man. Clearly an army doctor, then. He has just come from the tropics, for his face is dark, and that is not the natural tint of his skin, for his wrists are fair. He has undergone hardship and sickness, as his haggard face says clearly. His left arm has been injured. He holds it in a stiff and unnatural manner. Where in the tropics could an English army doctor have seen much hardship and got his arm wounded? Clearly in Afghanistan.

Whenever Holmes tells Watson such tales, Watson shakes his head in amazement. And we, Doyle's readers, are supposed to take our cue from Watson, thrilling to the detective's genius. But while Sherlock Holmes stories are good fun, it pays to notice that Holmes's method is ridiculous.

Take the rich story Holmes concocts after glancing at Watson in the lab. Watson is dressed in ordinary civilian clothes. What gives him "the air of a military man"? Watson is not carrying his medical bag or wearing a stethoscope around his neck. What identifies him as "a gentleman of a medical type"? And why is Holmes so sure that Watson had just returned from Afghanistan rather than from one of many other dangerous tropical garrisons where Britain, at the height of its empire, stationed troops? (Let's ignore the fact that Afghanistan is not actually in the tropical band.) And why does Holmes jump to the conclusion that Watson has sustained a battle wound? Watson holds his arm stiffly, but how does Holmes know that this isn't a result of a cricket injury? How does he know that Watson isn't experiencing—in his painful left arm—a classic symptom of a heart attack?

In short, Sherlock Holmes's usual method is to fabricate the most confident and complete explanatory stories from the most ambiguous clues. Holmes seizes on one of a hundred different interpretations of a clue and arbitrarily insists that the interpretation is correct. This then becomes the basis for a multitude of similarly improbable interpretations that all add up to a neat, ingenious, and vanishingly improbable explanatory story.

Sherlock Holmes is a literary figment. He lives in Neverland, so he always gets to be right. But if he tried to ply his trade as a "consulting detective" in the real world, he would be a dangerously incompetent boob—more like *The Pink Panther*'s Inspector Clouseau than the genius who lives with his friend Watson at 221b Baker Street.

We each have a little Sherlock Holmes in our brain. His job is to "reason backwards" from what we can observe in the present and show what orderly series of causes led to particular effects. Evolution has given us an "inner Holmes" because the world really is full of stories (intrigues, plots, alliances, relationships of cause and effect), and it pays to detect them. The storytelling mind is a crucial evolutionary adaptation. It allows us to experience our lives as coherent, orderly, and meaningful. It is what makes life more than a blooming, buzzing confusion.

But the storytelling mind is imperfect. After almost five decades of studying the tale-spinning homunculus who re- sides in the left brain, Michael Gazzaniga has concluded that this little man — for all of his undeniable virtues — can also be a bumbler. The storytelling mind is allergic to uncertainty, randomness, and coincidence. It is addicted to meaning. If the storytelling mind cannot find meaningful patterns in the world, it will try to impose them. In short, the storytelling mind is a factory that churns out true stories when it can, but will manufacture lies when it can't.

GEOMETRIC RAPE

The human mind is tuned to detect patterns, and it is biased toward false positives rather than false negatives. The same

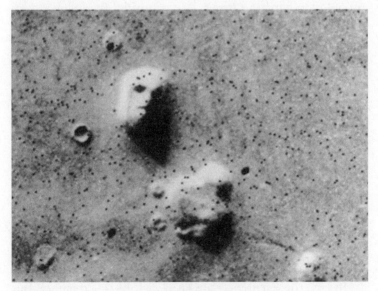

Image of a "face" on Mars taken by *Viking 1* in 1976. While some seized on the face as evidence of a Martian civilization, higher-resolution images showed that the "face" is just an ordinary Martian hill.

mental software that makes us very alert to human faces and figures causes us to see animals in clouds or Jesus in griddle marks. According to psychologists, this is part of a "mind design" that helps us perceive meaningful patterns in our environments.

Our hunger for meaningful patterns translates into a hunger for story. As the video game designer and writer James Wallis puts it, "Human beings like stories. Our brains have a natural affinity not only for enjoying narratives and learning from them but also for creating them. In the same way that your mind sees an abstract pattern and resolves it into a face, your imagination sees a pattern of events and resolves it into a story." There are a lot of neat studies that make Wallis's point, showing how we automatically extract stories from the information we receive, and how—if there is no story there—we are only too happy to invent one. Consider the following information:

Todd rushed to the store for flowers.

Greg walked her dog.

Sally stayed in bed all day.

Quick, what were you thinking? If you are like most people, you were puzzling over the three sentences, trying to find the hidden story. Perhaps Sally is sad because someone has died. Perhaps Greg and Todd are her friends: one is seeing to Sally's dog, and the other is buying her flowers. Or perhaps Sally is happy. She has just won the lottery, and to celebrate she has decided to luxuriate in bed all day. Greg and Todd are the underwear models she has hired as her masseurs and personal assistants.

In fact, these sentences are unrelated. I made them up. But if you have a healthy storytelling mind, you will automatically start to weave them together into the beginnings of a story. Of course, we recognize consciously that these sentences could serve as building blocks for an infinite number of narratives. But studies show that if you give people random, unpatterned information, they have a very limited ability *not* to weave it into a story.

This point is beautifully illustrated in an experiment by psychologists Fritz Heider and Marianne Simmel. In the mid-1940s, the researchers made a short animated film. The film is very simple. There is a big square that is motionless, except for a flap in one side that opens and closes. There is also a big triangle, a small triangle, and a small circle. The film opens with the big triangle inside the big square. The small triangle and the small circle then appear. As the big square flaps open and shut, the other geometric figures slide around the screen. After ninety seconds or so, the small triangle and the small circle disappear again.

When I first watched the film, I didn't see crude animated shapes moving randomly on a screen; I saw a weirdly powerful geometric allegory. The small triangle was the hero. The

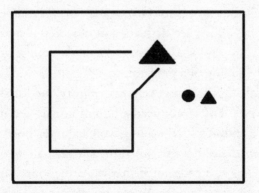

Re-creation of a screen shot of the Heider and Simmel film (1944), which can be viewed on YouTube. Other researchers have since replicated Heider and Simmel's findings many times.

big triangle was the villain. And the small circle was the heroine. The small triangle and the small circle enter the screen together, like a couple, and the big triangle storms out of his house (the square). The big triangle violently butts the little guy (small triangle) out of the way and herds the protesting heroine (small circle) back into his house. The big triangle then chases the circle back and forth, trying to work her into a corner. The scene reeks of sexual menace. Eventually the flap in the big square opens, and the small circle flees outside to join the small triangle. The couple (small triangle, small circle) then zip around the screen with the big triangle in hot pursuit. Finally, the happy couple escape, and the big triangle throws a tantrum and smashes his house apart.

It's a silly interpretation, of course. Like Sherlock Holmes, I've composed a rich and confident story from ambiguous clues. But I'm not alone in this. After showing this film to research subjects, Heider and Simmel gave them a simple task: "Describe what you saw." It's fascinating to note that only 3 of 114 subjects gave a truly reasonable answer. These people reported seeing geometric shapes moving around a screen, and that was all. But the rest of Heider and Simmel's subjects were like me; they didn't see fleshless and bloodless shapes sliding around. They saw soap operas: doors slamming, courtship dances, the foiling of a predator.

Similarly, in the early twentieth century, the Russian filmmaker Lev Kuleshov produced a film of unnarrated images: a corpse in a coffin, a lovely young woman, and a bowl of soup. In between these images, Kuleshov squeezed shots of an actor's face. The audience noted that when the soup was shown, the actor emoted hunger. When the corpse was shown, he looked sad. When the lovely young woman appeared, the actor's face was transformed by lust.

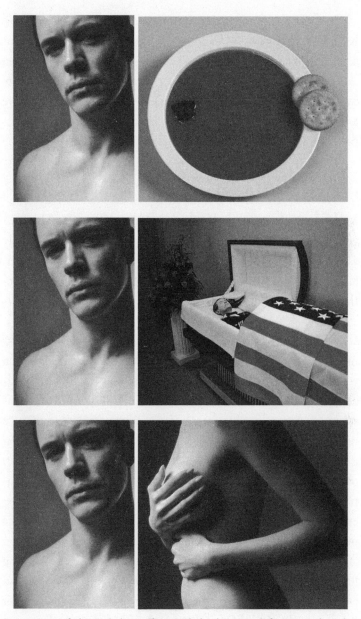

Re-creation of the Kuleshov effect. Kuleshov's original footage is lost, but many re-creations of his experiment can be viewed on YouTube.

In fact, the actor wasn't emoting at all. After every shot, Kuleshov had inserted exactly the same footage of an actor staring impassively into the camera. There was no hunger, sadness, or lust on the actor's face except what was put there by the audience. Kuleshov's exercise shows how unwilling we are to be without a story, and how avidly we will work to impose story structure on a meaningless montage.

IT'S JUST A FLESH WOUND

More than two hundred years ago, James Tilly Matthews invented an elaborate fictional world and lived stubbornly within it. His story is an example of pathological confabulation—of the creation of wild and largely implausible fictions that a person nonetheless believes with "rock-jawed certainty."

Pathological confabulators hew to their stories with striking tenacity. A brain-damaged father insists that he is still a young man, while also correctly listing the ages of his middle-aged children. Amnesiacs with Korsakoff's syndrome constantly forget who they are and constantly spin and re-spin new identities for themselves. As Oliver Sacks puts it, one patient with Korsakoff's is a "confabulatory genius" who "*must literally make himself (and his world) up every moment.*" Patients with Cotard's syndrome maintain a host of interesting explanations for why they appear to be alive when they are in fact stone dead. Confabulators who have lost an arm or a leg may adamantly deny it. When asked to move the missing limb, the amputee will invent a reason for not doing so, an explanation that involves arthritis or a rebellion against her doctor's badgering. (Then there is the Black Knight from the

1975 film *Monty Python and the Holy Grail*. After King Arthur cleaves away both of the knight's arms, the confabulating knight insists, "It's just a flesh wound." Arthur points out, "You've got no arms left!" As arterial blood spurts from his stumps, the knight responds, "Yes, I have!")

These examples of confabulation in damaged or diseased brains are not only fascinating in themselves; they are also fascinating for the way they illuminate the function of healthy minds. Psychologists are finding that ordinary, mentally healthy people are strikingly prone to confabulate in everyday situations. We're just so skillful that it's hard to catch us in the act. It takes clever lab work to show how often our storytelling minds run amok.

One of the first such studies was published by Norman Maier in 1931. Maier placed each of his research subjects in a room that was empty except for two ropes hanging from the ceiling and some objects lying about the room, such as an extension cord and a pair of pliers. The subject was given a job:

tie the two ropes together. But the ropes were too far apart to be grasped at the same time, and in many cases the subject was stumped. At some point, the psychologist would enter the room to check the subject's progress. The subject would then suddenly figure the problem out: tie the pliers to one of the ropes as a weight; get the weighted rope swinging; catch and tie.

When the psychologist asked the subjects how they hit upon the pliers method, they told great stories. One subject, who was himself a professor of psychology, responded, "I thought of the situation of swinging across a river. I had imagery of monkeys swinging from trees. This imagery appeared simultaneously with the solution. The idea appeared complete." What the subjects failed to mention was the event that had actually inspired the solution. When the researcher entered the room, he "accidentally on purpose" nudged one of the ropes and set it swinging, an event that few of the subjects consciously registered.

In a more recent study, psychologists asked a group of shoppers to choose among seven pairs of identically priced socks. After inspecting the socks and making their choices, the shoppers were asked to give reasons for their choices. Typically, shoppers explained their choices on the basis of subtle differences in color, texture, and quality of stitching. In fact, all seven pairs of socks were identical. There actually was a pattern in the shoppers' preferences, but no one was able to detect it: they tended to choose socks on the right side of the array. Instead of answering that they had no idea why they chose the socks they did, the shoppers told a story that made their decisions appear to be rational. But they weren't. The stories were confabulations—lies, honestly told.

A CURSED RAGE FOR ORDER

These may seem like tame examples. Does it really matter whether the stories we tell ourselves all day long—about why a spouse closed the laptop as we entered the room or why a colleague has a guilty look on her face—have a basis in fact? Who cares whether we make up stories about socks? But there is a dark side to our tendency to impose stories where they do not exist, and nothing reveals it like a good conspiracy theory.

Conspiracy theories—feverishly creative, lovingly plotted—are in fact fictional stories that some people believe. Conspiracy theorists connect real data points and imagined data points into a coherent, emotionally satisfying version of reality. Conspiracy theories exert a powerful hold on the human imagination—yes, perhaps even *your* imagination—not *despite* structural parallels with fiction, but in large part *because of* them. They fascinate us because they are ripping good yarns, showcasing classic problem structure and sharply defined good guys and villains. They offer vivid, lurid plots that translate with telling ease into wildly popular entertainment. Consider novels such as Dan Brown's *The Da Vinci Code* and James Ellroy's Underworld USA Trilogy, films such as *JFK* and *The Manchurian Candidate,* and television shows such as *24* and *The X-Files.*

Loud, angry, and charismatic, Alex Jones has built a commercial mini-empire by peddling stories of evil conspiracies. In his documentary *Endgame: Blueprint for Global Enslavement,* Jones—chubby and pushing forty—offers a river of innuendo and a dearth of actual evidence to suggest that a tiny cadre of "global elites" are executing a dastardly plan to take over the earth and enslave its inhabitants. Road signs have

Alex Jones leading a 9/11 Truth rally in New York City.

been specially marked so that invading UN troops will be able to navigate the American heartland. Mailboxes have been discreetly marked, too. If you find a discreet blue dot on yours, breathe a sigh of relief: you will only be sent to a FEMA concentration camp. If you find a red dot, say your prayers: the foreign invaders will execute you on the spot. According to Jones, the plotters, Lucifer worshipers to a man, will depopulate the world by 80 percent and then take advantage of advances in medical genetics that will allow them to live forever like gods.

Documentarians from the Independent Film Channel (IFC) recently followed Jones around the country as he roared against the New World Order, investigated the assassination of JFK, and fired up the crowds at 9/11 Truth rallies. In the IFC documentary and on his disturbingly popular radio program (it draws one million listeners per day), Jones comes off as a paranoid egomaniac. Wherever he goes, he is sure he is being followed. Driving down the highway, he and his colleagues are constantly swiveling their heads, alert for tails. When an ordinary sedan passes them, Jones snaps pictures of the car, saying, "Oh, yeah, that's military intelligence." Standing outside the White House, Jones is convinced that a shirtless mountain biker is an undercover Secret Service agent keeping tabs on him. When a fire alarm goes off in his hotel just as he is about to be interviewed on a live radio program, he spits and sputters and flaps his arms, roaring, "They've set us up! They've set us up!"

It's fair to ask what, if anything, separates the paranoid confabulations of Alex Jones from those of James Tilly Matthews—or from the influential conspiracy theorist David Icke, who appears not to be kidding when he argues that the world is ruled by vampiristic extraterrestrial lizard people in disguise. It seems that differences between the delusions of psychotics and the fantasies of conspiracy theorists are of degree, not kind. Matthews's delusional psychosis appears to be a broken version of the same pattern hunger that gives us conspiracy theories in mentally healthy individuals. In conspiracy theories, we have the storytelling mind operating at its glorious worst.

What's really striking about conspiracy theories is not their strangeness but their ordinariness. Go to Google, type in "conspiracy," and browse through some of the 37 million hits. You will find that there is a conspiracy theory for just about everything. There are the big classics, invoking evil cabals of Illuminati, Masons, and Jews. There is a conspiracy theory for any major entertainment or political figure who dies young: Marilyn, Elvis, Biggie, and Tupac; Princess Di (murdered be-

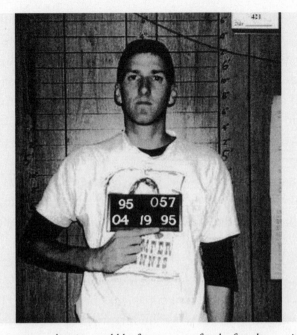

Many conspiracy theories would be funny except for the fact that stories—no matter how fanciful—have consequences. For example, in Africa many believe that AIDS is a racist hoax designed to terrify black people into abstinence and condom use, and thus to perpetrate a bloodless genocide. Believing this gets a lot of Africans killed. Timothy McVeigh (pictured here), who has been described as "a walking compendium of antigovernment conspiracy theories," blew up the federal building in Oklahoma City because he subscribed to a prime tenet of the militia movement: the U.S. government has sold out to the New World Order.

cause she had an Arab baby in her womb); RFK, JFK, and MLK (all killed by the same Manchurian candidate). There are conspiracy theories about Hurricane Katrina (government operatives dynamited the levees to drown black neighborhoods), fluoridated drinking water (a means of mind control), aphrodisiac bubble gum (Israelis use it to turn Palestinian girls into tarts), jet plane vapor trails (they spew aggression-enhancing chemicals into minority neighborhoods), Paul Mc-

Cartney (long dead), John Lennon (gunned down by Stephen King), the Holocaust (didn't happen), Area 51 cover-ups (happened), moon landings (didn't), and so on.

A truly stunning number of people actually believe these stories. For example, a July 2006 Scripps Howard poll showed that 36 percent of Americans thought the U.S. government was complicit in the 9/11 attacks, with a majority of Democrats and young people (between the ages of eighteen and thirty) believing that government elements either actively pulled off the attacks or—in a rehash of Pearl Harbor conspiracy theories—had foreknowledge of the attacks but did nothing to stop them. The left certainly hasn't cornered the market on conspiracy. At the time of this writing, for example, large numbers of right-wing Americans are lost in fantasies about President Barack Obama. Obama is a stealth Muslim (one-third of conservative Republicans believed this as of August 2010, along with 20 to 25 percent of Americans generally). Obama was not born in the United States (45 percent of Republicans). Obama is a communist who is actively trying to destroy America. Obama wants to set up Nazi-style death panels to euthanize old people. Obama is the Antichrist (in a controversial Harris Poll, 24 percent of Republicans endorsed the statement that Obama "might" be the Antichrist).

It's tempting to blame general backwardness or ignorance for this epidemic of conspiracy. Tempting, but wrong. As David Aaronovitch explains in his book *Voodoo Histories*,

> Conspiracy theories originate and are largely circulated among the educated and middle class. The imagined model of an ignorant, priest-ridden peasantry or proletariat, replacing religious and superstitious belief with equally far-fetched notions of how

society works, turns out to be completely wrong. It has typically been the professors, the university students, the managers, the journalists, and the civil servants who have concocted and disseminated the conspiracies.

Conspiracy theories are not, then, the province of a googly-eyed lunatic fringe. Conspiratorial thinking is not limited to the stupid, the ignorant, or the crazy. It is a reflex of the storytelling mind's compulsive need for meaningful experience. Conspiracy theories offer ultimate answers to a great mystery of the human condition: why are things so bad in the world? They provide nothing less than a solution to the problem of evil. In the imaginative world of the conspiracy theorist, there is no accidental badness. To the conspiratorial mind, shit *never* just happens. History is *not* just one damned thing after another, and only dopes and sheeple believe in coincidences. For this reason, conspiracy theories—no matter how many devils they invoke—are always consoling in their simplicity. Bad things do not happen because of a wildly complex swirl of abstract historical and social variables. They happen because bad men live to stalk our happiness. And you can fight, and possibly even defeat, bad men. *If* you can read the hidden story.

The Moral of the Story

We live or die by the artist's vision, sane or cracked.

— JOHN GARDNER, *On Moral Fiction*

FLIP THROUGH THE holy books of the three great monotheisms—Judaism, Christianity, and Islam—and you will be flipping through anthologies of stories: the Fall, the Flood, Sodom and Gomorrah, Abraham and Isaac, the crucifixion and resurrection of Christ, the Archangel Gabriel seizing Muhammad by the throat and revealing that Allah created man from a clot of blood. Take away the lists of begettings, the strings of "thou shalts" and "thou shalt nots" (one writer puts the number of biblical commandments not at ten, but at more like seven hundred), the instructions on how to sacrifice animals and how to build an ark, and you have a collection of intense narratives about the biggest stuff in human life. The Middle Eastern holy books are catalogs of savage violence, of a cruel God wantonly smiting, of a merciful God blessing and forgiving, of people suffering on the move, of men and women joining in love and doing lots and lots of begetting.

And, of course, it is not just the planet's monotheisms that are built on stories. This seems to be true of all religions, major and minor, throughout world history. Read through the folklore of traditional peoples, and the dominant story type will be myths explaining why things are the way they are. In traditional societies, truths about the spirit world were conveyed not through lists or essays — they were conveyed through story. The world's priests and shamans knew what psychology would later confirm: if you want a message to burrow into a human mind, work it into a story.

Staunch believers in any of the three major monotheisms (Christianity, Islam, and Judaism) may take offense when I refer to their holy scriptures as stories. But many of those same believers would be quick to say that narratives about Zeus or Thor or Shiva — the Hindu god of destruction (pictured here) — are just stories.

Guided by the holy myths, believers must imaginatively construct an alternate reality that stretches from the origins straight through to the end times. Believers must mentally simulate an entire shadow world that teems with evidence of divinity. They must be able to decode the cryptic messages in the stars, the whistle of the wind, the entrails of goats, and the riddles of the prophets.

Throughout the history of our species, sacred fiction has dominated human existence like nothing else. Religion is the ultimate expression of story's dominion over our minds. The heroes of sacred fiction do not respect the barrier between the pretend and the real. They swarm through the real world, exerting massive influence. Based on what the sacred stories say, believers regulate the practices of their lives: how they eat, how they wash, how they dress, when they have sex, when they forgive, and when they wage total war in the name of everything holy.

Why?

Religion is a human universal, present—in one form or another—in all of the societies that anthropologists have visited and archaeologists have dug up. Even now, in this brave age of brain science and genomics, God is not dead, dying, or really even ailing. Nietzsche would be disappointed. Most of the world's people don't look up at the sky and find it—like the poet Hart Crane—"ungoded." The world's big religions are gaining more converts than they are losing. While Europe has become more secular over the past century, most of the rest of the world (including the United States) is getting more religious.

Since it is not plausible that religion just happened to develop independently in many thousands of different cultures, *Homo sapiens* must have already been a spiritual ape when our ancestors began streaming out of Africa. And since all religions share some of the same basic features—including belief in supernatural beings, belief in a transcendent soul, belief in the efficacy of magic (in the form of rituals and prayer)—the roots of spirituality must be sunk deep in human nature.

But why did we evolve to be religious? How did dogmatic faith in imaginary beings *not* diminish our ability to survive and reproduce? How could the frugal mechanisms of natural selection *not* have worked against religion, given the high price of religious sacrifices, rituals, prohibitions, taboos, and commandments? After all, burning a goat for Zeus meant one less goat for your family. And sawing off a perfectly good piece of your baby son's penis because an ancient story suggests you should is not without risks. (Before the discovery of the germ theory of disease—in the days before antibiotics and surgical steel—circumcision was dangerous, and accidents still do happen.) Also, while it is pretty easy to refrain from violating biblical injunctions against wearing cloth with mixed fibers or boiling a baby goat in the milk of its mother, it's more of a burden to be stoning people all the time: adulterers, magicians, Sabbath breakers, incest enthusiasts, blasphemers, disobedient children, idolaters, wayward oxen.

Religious tendencies are either an evolutionary adaptation, an evolutionary side effect, or some combination of the two. The conventional secular explanation of religion is that humans invent gods to give order and meaning to existence. Humans are born curious, and they must have answers to the big, unanswerable questions: Why am I here? Who made me? Where does the sun go at night? Why does giving birth hurt? What happens to "me" after I die—not my raggedy old carcass, but *me,* that endlessly chattering presence inside my skull?

This is, in essence, a by-product explanation of religion,

and it is the one that most current evolutionary thinkers embrace. We have religion because, by nature, we abhor explanatory vacuums. In sacred fiction, we find the master confabulations of the storytelling mind.

Some evolutionary thinkers, including leading lights such as Daniel Dennett and Richard Dawkins, focus relentlessly on the black side of religious behavior: the pogroms, the bigotry, the suppression of real thought in favor of dumb faith. They think that religion is the result of a tragic evolutionary glitch. Dawkins and Dennett argue that the mind is vulnerable to religion in the same way that a computer is vulnerable to viruses. Both Dennett and Dawkins view religion as a mental parasite (as Dawkins memorably put it, religion is "a virus of the mind"), and a noxious one at that. For these thinkers, religion is not akin, say, to the friendly parasites that colonize our intestines and help us digest food. Religion is more like the loathsome pinworms that lay itchy eggs around the anus. According to Dawkins and Dennett, human life would be a lot better if the mental parasite of religion could simply be eradicated.

I'm not so sure. I think the by-product explanation of religion captures a major part of the truth: humans conjure gods, spirits, and sprites to fill explanatory voids. (This is not to deny the possibility of gods, spirits, or sprites; it is to deny that one culture's supernatural story can be more valid than another's.) But does this mean that religion is, in evolutionary terms, useless or worse? A growing number of evolutionists think not.

In his trailblazing book *Darwin's Cathedral*, the biologist David Sloan Wilson proposes that religion emerged as a stable part of all human societies for a simple reason: it made them work better. Human groups that happened to possess a

faith instinct so thoroughly dominated nonreligious competitors that religious tendencies became deeply entrenched in
our species.

Wilson argues that religion provides multiple benefits to
groups. First, it defines a group as a group. As the sociologist
Émile Durkheim wrote, "Religion is a unified system of beliefs and practices . . . which unite into one single moral community called a Church all those who adhere to them." Second, religion coordinates behavior within the group, setting
up rules and norms, punishments and rewards. Third, religion
provides a powerful incentive system that promotes group co-
operation and suppresses selfishness. The science writer Nicholas Wade expresses the heart of Wilson's idea succinctly: the
evolutionary function of religion "is to bind people together
and make them put the group's interests ahead of their own."

Atheists are often dismayed that intelligent believers can
entertain patently irrational beliefs. From the atheist's perspective, the earth's faithful are like billions of foolish Don
Quixotes jousting with windmills—all because, like Quixote,
they can't see that their favorite storybooks are exactly that.

But Wilson points out that "elements of religion that appear irrational and dysfunctional often make perfectly good
sense when judged by the only appropriate gold standard as
far as evolutionary theory is concerned—*what they cause people to do.*" And what they generally cause people to do is to behave more decently toward members of the group (coreligionists) while vigorously asserting the group's interests against
competitors. As the German evolutionist Gustav Jager argued
in 1869, religion can be seen as "a weapon in the [Darwinian]
struggle for survival."

As Jager's language suggests, none of this should be con-
strued to suggest that religion is—on the whole—a good

thing. There are good things about religion, including the way its ethical teachings bind people into more harmonious collectives. But there is an obvious dark side to religion, too: the way it is so readily weaponized. Religion draws coreligionists together, *and* it drives those of different faiths apart.

SACRED HISTORIES

Supernatural myths aren't the only stories that play a binding role in society. National myths can serve the same function. I recently asked my first-grade daughter, Abigail, to tell me what she learned in school about Christopher Columbus. Abby has an excellent memory, and she recalled a lot: the names of the three ships, the fact that Columbus discovered America by sailing the ocean blue in 1492, and that Columbus proved that the earth was round, not flat. It's the same thing they taught me in elementary school thirty years ago, and what my parents learned before me.

A depiction of Columbus arriving in the New World by Dióscoro Puebla (1831–1901).

But what Abigail was taught is mostly fiction, not history. It is a story that is simply wrong in most details and misleading in the rest. On the small side of things, in 1492 most educated people shared Columbus's confidence that the earth was round. On the large side of things, Abby was not told that Columbus first landed in the West Indies, where he wrote of the Arawak Indians, "They would make fine servants . . . With fifty men we could subjugate them all and make them do whatever we want." Columbus and his followers went on to do just that, killing and enslaving the Arawak with real avidity and sadistic creativity. Within sixty years or so, the Arawak were wiped out. Abby also wasn't told that this was just the first stage of a centuries-long effort to strip the North American continent of Indian life.

Revisionist historians such as Howard Zinn and James Loewen have argued that American history texts have been whitewashed so thoroughly that they don't count as history anymore. They represent determined forgetting—an erasure of what is shameful from our national memory banks so that history can function as a unifying, patriotic myth. Stories about Columbus, Squanto and the first Thanksgiving, George Washington's inability to lie, and so on, serve as national creation myths. The men at the center of these stories are presented not as flesh-and-blood humans with flaws to match their virtues, but as the airbrushed leading men of hero stories. The purpose of these myths is not to provide an objective account of what happened. It is to tell a story that binds a community together—to take *pluribus* and make *unum*.

Many commentators see revisionists like Zinn and Loewen not as myth busters, but as spinners of countermyths, in which Western society is trashed and indigenous societies are absurdly romanticized. They point out that societies everywhere—including the New World and African societies decimated by Westerners—have long histories of war and conquest. For them, the big difference between the conquering Western powers and their victims was technological. For example, if the rapacious Aztec Empire had developed the means to sail to Europe in order to sack and pillage, they may well have done so. People have been pretty nasty throughout history, and over the past half millennium or so, Westerners have just been better at being nasty than anyone else.

But this more balanced, if bleak, view of human history isn't taught in our schools either. Throughout most of our history, we've taught myths. The myths tell us that not only are we the good guys, but we are the smartest, boldest, best guys that ever were.

IMAGINING THE UNIMAGINABLE

Theirs was a May-December affair. Tom was only twenty-two. He was tall and lean, boyish of face and build. Sarah was buxom and quick to laugh. She looked much younger than her forty-five years. If not for the silver streaks in her dark hair, she might have passed for Tom's sister. When Tom graduated college, Sarah decided to take him to Paris as a reward. "Let me be your sugar momma," she said, laughing.

They spent ten days in the city, gawking at the Eiffel Tower, the wonders of the Louvre, and the massive spectacle of Notre Dame. On their next-to-last night, they ate dinner and drank red wine at an almost absurdly charming bistro in the Latin Quarter. Tom and Sarah noticed the other patrons' stares. People were always staring. As they strolled hand in hand down Parisian boulevards, they felt the strangers' eyes appraising them, judging them, tsk-tsking behind their backs.

They knew what people were thinking: that it was not right, that she was old enough to be his mother.

Maybe the Parisians weren't thinking this at all. Maybe the Parisians were just staring because they were an attractive couple and clearly much in love. Maybe Tom and Sarah were paranoid. But if so, they had reason to be. The lovers had paid dearly for their bliss. After learning of the affair, Sarah's mother had vowed not to speak to her until she got therapy. Sarah's coworkers whispered nastily about her in the office lounge. For his part, Tom had tearful fights with his father over the affair. When he told his fraternity brothers about Sarah, they laughed nervously, thinking it was a bad joke. But when Sarah started spending nights in Tom's room, the brothers held an emergency meeting and tried to expel Tom from the house.

Truth be told, the lovers enjoyed being judged. That was half the fun of it. They liked thinking of themselves as rebels with the courage to live in contempt of convention. They sat in the bistro finishing their wine and asking each other their favorite rhetorical questions: Who were they harming? Why were people so nosy and jealous? Why was their morality so narrow and timid?

They walked back to their hotel along the Seine, drunk on wine and rebellion. Entering their room, Tom hung the NE PAS DÉRANGER sign on the doorknob. Then, bouncing and rolling across the room's surfaces, Tom and Sarah made love with an athletic intensity bordering on violence.

Afterward, Tom collapsed on Sarah. She cuddled his panting head to her breast and stroked his curly hair and murmured sweetly in his ear. When Tom regained his breath, he rolled to his own pillow and said, "So what should we see tomorrow, Mom?"

Sarah lay her head on his shoulder and plucked playfully at the sparse hairs on his chest. Giggling, she tickled him. "Maybe we'll just stay in bed tomorrow!" Tom giggled back at her, saying, *"Mo-om! Sto-op!"*

First, I'm sorry I did that to you—questionable taste, I agree. Second, I have my reasons, which I'll get to now.

If you are like most people, you probably couldn't help imagining this fictional Parisian love affair. If you have visited Paris, images of the famous city invaded your brain. If you have never visited the city, views from movies, paintings, and postcards still crowded into your head, along with generic images of lovers enjoying a meal in a charming restaurant. When the action turned to joyful lovemaking between two attractive people, your interest in the story might have picked up.

You might have imagined what Tom and Sarah looked like unclothed and how they moved together. After all, such moments—or the promise of them—may be *the* main staple not just of romance novels and porn but of all story types.

But what happened when you learned that the lovers were mother and son? Did your mind revolt? Did you find yourself trying to expel the images from your brain, the way a child might spit out a bite of cake that tasted delicious until she learned it was made from hated carrots?

My loathsome little story was inspired by the psychologist Jonathan Haidt, who exposes people to uncomfortable fictional scenarios like this one in order to study moral logic. (In addition to consensual incest, Haidt's uncomfortable scenarios include a man who has victimless sex with dead chickens and a family that devours their beloved dog after he is killed by a car.) If the man and woman in this story were not related, you probably would have enjoyed imagining their lovemaking. But knowing the truth sours the fantasy. Even if the story clearly shows that the two are consenting adults enjoying an emotionally and physically satisfying relationship, most people are unwilling even to imagine that the relationship is morally acceptable.

This is remarkable, because people are willing to imagine almost anything in a story: that wolves can blow down houses; that a man can become a vile cockroach in his sleep (Franz Kafka's "The Metamorphosis"); that donkeys can fly, speak, and sing R&B songs (*Shrek*); that "a dead-but-living fatherless god-man [Jesus] has the super-powers to grant utopian immortality"; that a white whale might really be evil incarnate; that time travelers can visit the past, kill a butterfly, and lay the future waste (Ray Bradbury's "A Sound of Thunder").

I should say that people are willing to imagine *almost* anything. This flexibility does not extend to the moral realm. Shrewd thinkers going back as far as the philosopher David Hume have noted a tendency toward "imaginative resistance": we won't go along if someone tries to tell us that bad is good, and good is bad.

Here's how Dostoyevsky *didn't* write *Crime and Punishment:* Raskolnikov kills the pawnbroker and her sister for kicks; he feels no remorse; he boasts to his family and friends about it, and they all wet themselves laughing; Raskolnikov is a good man; he lives happily ever after.

Or imagine a short story based on Jonathan Swift's satirical essay "A Modest Proposal." Swift suggests that all of the social ills of Ireland can be solved by starting a baby-meat industry. But imagine that the story is not a satire. Imagine a story in which the author fails to signal that it would be wrong for impoverished women to fatten infants at their breasts just so rich men could feast on baby chops and baby ragout.

Or imagine a story based on the Roman emperor Heliogabalus, who is said to have slaughtered slaves on his front lawn—men, women, and children—just because he found the shimmer of blood on grass delightful to gaze upon. The story celebrates Heliogabalus as an artistic pioneer who fearlessly pursued beauty in the face of arbitrary moral codes.

In the same way that we are unwilling to imagine a scenario in which it is okay for a mother and son to be lovers, most of us are unwilling to imagine a universe where the murder of slaves, babies, or pawnbrokers is morally acceptable.

Storytellers know this in their blood. True, they deluge us with breathtaking depravity, lewdness, and cruelty—think *Lolita,* think *A Clockwork Orange,* think *Titus Andronicus* (in this Shakespeare play, "two men kill another man, rape his bride, cut out her tongue, and amputate her hands; her father kills the rapists, cooks them in a pie, and feeds them to their mother, whom he then kills before killing his own daughter for having gotten raped in the first place; then he is killed, and his killer is killed"). And we love them for it. We are only too happy to leer on as the bad guys of fiction torture, kill, and rape. But storytellers never ask us to approve. Morally repellent acts are a great staple of fiction, but so is the storyteller's condemnation. It was very wrong, Dostoyevsky makes clear, for Raskolnikov to kill those women. It would be very wrong, Swift makes clear, to raise babies like veal, no matter the socioeconomic returns.

VIRTUE REWARDED

The Greek philosopher Plato banished poets and storytellers from his ideal republic for, among other sins, peddling immoral fare. And Plato's was just the first in a long string of panic attacks about the way fiction corrodes morality—how penny dreadfuls, dime novels, comics, moving pictures, television, or video games are corrupting the youth, turning them slothful and aggressive and perverted.

But Plato was wrong, and so were his panicked descendants. Fiction is, on the whole, intensely moralistic. Yes, evil occurs, and antiheroes, from Milton's Satan to Tony Soprano, captivate us. But fiction virtually always puts us in a position to judge wrongdoing, and we do so with gusto. Sometimes we find ourselves rooting perversely for dark heroes such as Satan or Soprano, or even the child molester Humbert Humbert in *Lolita,* but we aren't asked to approve of their cruel and selfish behavior, and storytellers almost never allow them to live happily ever after.

Madame Bovary disrobing for her lover, Léon. In 1857, Gustave Flaubert was tried on the charge that *Madame Bovary* was an outrage against morality and religion. Flaubert's lawyer successfully argued that although the novel depicts immoral acts, it is itself moral. Emma Bovary sins, and she suffers for it.

One of the first novels in English was Samuel Richardson's *Pamela; or, Virtue Rewarded* (1740). That subtitle could be tacked onto most stories that humans have dreamed up, from the first folktales to modern soap operas and professional wrestling. Story runs on poetic justice, or at least on our hopes for it. As the literary scholar William Flesch shows in his book *Comeuppance,* much of the emotion generated by a story—the fear, hope, and suspense—reflects our concern over whether the characters, good and bad, will get what they deserve. Mostly they do, but sometimes they don't. And when they don't, we close our books with a sigh, or trudge away from the theater, knowing that we have just experienced a tragedy.

By the time American children reach adulthood, they will have seen 200,000 violent acts, including 40,000 killings, on television alone—which is to say nothing of film or the countless enemies they have personally slaughtered in video games. Social scientists generally frown at this carnage, arguing that it leads to an increase in real-world aggression. They have a point (which we'll examine more closely in the next chapter). But they also miss one. Fiction almost never gives us morally neutral presentations of violence. When the villain kills, his or her violence is condemned. When the hero kills, he or she does so righteously. Fiction drives home the message that violence is acceptable only under clearly defined circumstances—to protect the good and the weak from the bad and the strong. Yes, some video games, such as Grand Theft Auto, glorify wickedness, but those games are the notorious exceptions that prove the general rule.

The psychologist Jerome Bruner writes that "great fiction is subversive in spirit." I disagree. It's true that writers have frequently, especially over the past century, set out to challenge (or outrage) conventional sensibilities. There is a reason for all those burned and banned books. But most of this fiction is still moral fiction: it puts us in the position of approving of decent, prosocial behavior and disapproving of the greed of antagonists—of characters who are all belly and balls. As novelists such as Leo Tolstoy and John Gardner have argued, fiction is, in its essence, deeply moral. Beneath all of its brilliance, fiction tends to preach, and its sermons are usually fairly conventional.

In Charles Baxter's influential book on the craft of fiction,

A STORY-TELLER RECITING FROM THE "ARABIAN NIGHTS."

An Egyptian woman telling tales from "Arabian Nights." It's worth remembering that until recently, storytellers who attacked group values faced real risks. For tens of thousands of years before the invention of the book, story was an exclusively oral medium. Members of a tribe gathered around a teller and listened. Tribal storytellers who undercut time-honored values—who insulted group norms—faced severe consequences. (Imagine if our Egyptian storyteller decided to spin a yarn about the wine-soaked debauches of the Prophet, Muhammad.) As a result, oral stories generally reflect "a highly traditionalist or conservative set of mind."

Burning Down the House, he bemoans the "death of the antagonist—any antagonist" in modern fiction. He's onto something. Over the past hundred years or so, sophisticated fiction has trended toward moral ambiguity. This is strikingly illustrated by the edgy protagonists of recent cable TV dramas such as *The Shield, The Wire, Dexter, Breaking Bad, The Sopranos,* and *Deadwood.* But I'm making a general argument, not an absolute one. I think that rumors of the death of the antagonist have been exaggerated. Take those edgy antiheroes from cable drama. Do they really muddle the ethical patterns I'm describing or—by setting virtue against vice *inside* the soul of Walter White or Tony Soprano—do they just put a fresh twist on old morality plays? In any case, I agree with the journalist Steven Johnson, who concludes that the most popular story forms—mainstream films, network television, video games, and genre novels—are still structured on poetic justice: "the good guys still win out, and they do it by being honest and playing by the rules."

If there really is a general pattern of conventional moralizing in stories—one that stands out around the world despite some exceptions—where does it come from? William Flesch thinks it reflects a moralistic impulse that is part of human nature. I think he's right. I think it reflects this impulse, but I also think it *reinforces* it. In the same way that problem structure points up a potentially important biological function of story (problem rehearsal), the moralism of fiction may point up another important function.

In a series of papers and a forthcoming book, Joseph Carroll, John Johnson, Dan Kruger, and I propose that stories make societies work better by encouraging us to behave ethically. As with sacred myths, ordinary stories—from TV shows to fairy tales—steep us all in the same powerful norms and values. They relentlessly stigmatize antisocial behavior and just as relentlessly celebrate prosocial behavior. We learn by association that if we are more like protagonists, we will be more apt to reap the typical rewards of protagonists (for instance, love, social advancement, and other happy endings) and less likely to reap the rewards of antagonists (for instance, death and disastrous loss of social standing).

Humans live great chunks of their lives inside fictional stories—in worlds where goodness is generally endorsed and rewarded and badness is condemned and punished. These patterns don't just reflect a moralistic bias in human psychology, they seem to reinforce it. In his book *The Moral Laboratory*,

Evidence that poetic justice is basic to the fiction impulse comes from children's pretend play. According to David Elkind's *The Power of Play,* children's
pretend play always has clear "moral overtones—the good guys versus the bad
guys." Children's play scenarios are endlessly convulsed by the collision of evil
and good, as in this photograph of children playing cops and robbers.

the Dutch scholar Jèmeljan Hakemulder reviewed dozens of
scientific studies indicating that fiction has positive effects on
readers' moral development and sense of empathy. In other
words, when it comes to moral law, Shelley seems to have
had it right: "Poets are the unacknowledged legislators of the
world."

Similar evidence comes from a 2008 study of television
viewers by the psychologist Markus Appel. Think about it:
for a society to function properly, people have to believe in
justice. They have to believe that there are rewards for doing
right and punishments for doing wrong. And, indeed, people generally do believe that life punishes the vicious and rewards the virtuous. This is despite the fact that, as Appel puts

it, "this is patently not the case." Bad things happen to good people all the time, and most crimes go unpunished.

In Appel's study, people who mainly watched drama and comedy on TV—as opposed to heavy viewers of news programs and documentaries—had substantially stronger "just-world" beliefs. Appel concludes that fiction, by constantly marinating our brains in the theme of poetic justice, may be partly responsible for the overly optimistic sense that the world is, on the whole, a just place. And yet the fact that we take this lesson to heart may be an important part of what makes human societies work.

Go into a movie theater. Sit in the front row, but don't watch the movie. Turn around and watch the people. In the flickering light, you will see a swarm of faces—light and dark, male and female, old and young—all staring at the screen. If the movie is good, the people will respond to it like a single organism. They will flinch together, gasp together, roar with laughter together, choke up together. A film takes a motley association of strangers and syncs them up. It choreographs how they feel and what they think, how fast their hearts beat, how hard they breathe, and how much they perspire. A film melds minds. It imposes emotional and psychic unity. Until the lights come up and the credits roll, a film makes people one.

It has always been so. It is easy for us to forget, sitting alone on our couches with our novels and television shows, that until the past few centuries, story was always an intensely communal activity. For tens of thousands of years before the invention of writing, story happened only when a teller came together with listeners. It wasn't until the invention of the printing press that books became cheap enough to reward

mass literacy. For uncounted millennia, story was exclusively oral. A teller or actor attracted an audience, synched them up mentally and emotionally, and exposed them all to the same message.

In recent centuries, technology has changed the communal nature of story, but it has not destroyed it. Nowadays we may imbibe most of our stories alone or with our families and friends, but we are still engaged in a socially regulating activity. I may be by myself watching *Breaking Bad* or *30 Rock,* or reading *The Da Vinci Code* or *The Girl with the Dragon Tattoo,* but there are millions of other people sitting on millions of other couches being exposed to exactly the same stories and undergoing exactly the same process of neural, emotional, and physiological attunement. We are still having a communal experience; it's just spread out over space and time.

Story, in other words, continues to fulfill its ancient function of binding society by reinforcing a set of common values and strengthening the ties of common culture. Story en-

culturates the youth. It defines the people. It tells us what is
laudable and what is contemptible. It subtly and constantly
encourages us to be decent instead of decadent. Story is the
grease and glue of society: by encouraging us to behave well,
story reduces social friction while uniting people around common values. Story homogenizes us; it makes us one. This is
part of what Marshall McLuhan had in mind with his idea of
the global village. Technology has saturated widely dispersed
people with the same media and made them into citizens of a
village that spans the world.

Story—sacred and profane—is perhaps *the* main cohering
force in human life. A society is composed of fractious people with different personalities, goals, and agendas. What con-
nects us beyond our kinship ties? Story. As John Gardner puts
it, fiction "is essentially serious and beneficial, a game played
against chaos and death, against entropy." Story is the coun-
terforce to social disorder, the tendency of things to fall apart.
Story is the center without which the rest cannot hold.

Ink People Change the World

We are absurdly accustomed to the miracle of a few written signs being able to contain immortal imagery, involutions of thought, new worlds with live people, speaking, weeping, laughing.

— VLADIMIR NABOKOV, *Pale Fire*

ALOIS SCHICKLGRUBER WAS BORN in 1837 in the tiny village of Strones, in the hilly region north of Vienna. The Schicklgrubers were peasants, but Alois rose by pluck to a good job in the civil service. Alois raised his family in the town of Linz. He sired nine children in all, including a son named Adolfus, who lived for opera.

Adolfus's boyhood friend August Kubizek relates how, when Adolfus was just sixteen, the two boys attended a performance of Richard Wagner's opera *Rienzi*. For five full hours, the two friends gazed down from the cheap seats as the story of Cola Rienzi, the heroic Roman tribune of the people, unfolded in blasts of song. Afterward, exhausted and emotionally spent, the two friends walked the winding streets of Linz.

The voluble Adolfus was oddly quiet. In silence, he led

Adolfus as a baby.

his friend up the Freinberg, a hill overlooking the Danube. There Adolfus stopped and grasped Kubizek's hands. Trembling with "complete ecstasy and rapture," he said that *Rienzi* had revealed his destiny. "He conjured up in grandiose, inspiring pictures his own future and that of his people . . . He was talking of a mandate which, one day, he would receive from the people, to lead them out of servitude to the heights of freedom." Then Kubizek watched Adolfus walk away into the night.

As a young man, Adolfus dreamed of being a great painter. He dropped out of school at seventeen and moved to Vienna, hoping to attend the Academy of Fine Arts. But while Adolfus could paint landscapes and architectural scenes, he was defeated by the human form, and so he was twice rejected by the academy. Depressed, Adolfus slipped into an aimless, loafing existence. He whipped off paintings of Viennese landmarks and sold them to tourists for the equivalent, in today's money, of ten or fifteen dollars. He flopped for a time with winos and hoboes in a homeless shelter, walking the streets for hours to escape the bugs in his room. He took his meals in soup kitchens. In the winter, he shoveled snow to earn money and spent time in public warming rooms. He sometimes hung around the train station, carrying bags for tips.

Adolfus's relatives tried to get him jobs as a baker's apprentice and a customs officer. He brushed them off. Through all the years of struggle and failure, the confidence he gained from his *Rienzi* epiphany never wavered; he knew he would make his mark.

Adolfus's last name was not Schicklgruber. His father, Alois, had been born out of wedlock, so Alois was given his mother's last name. But Alois's mother later married Johann Georg Hiedler. When Alois was thirty-nine, he legally took his stepfather's name, which was variously spelled Hiedler, Huetler, or Hitler. The government clerk processing the name change settled on the last spelling, and Alois Schicklgruber became Alois Hitler.

One of Adolf ("Adolfus" is the name on his birth certificate) Hitler's best biographers, Ian Kershaw, writes, "Hitler is one of the few individuals of whom it can be said with absolute certainty: without him, the course of history would have been different." Historians have, therefore, speculated endlessly about whether the twentieth century might have taken a gentler turn if Hitler had been admitted to art school, or if he had not attended *Rienzi* that night in 1906 and gotten drunk on a fantasy of himself as his nation's savior.

Historians question much in August Kubizek's memoir, *The Young Hitler I Knew*, which began as a work of hero worship commissioned by the Nazi Party but was not finished until after World War II. However, the *Rienzi* episode seems

The Courtyard of the Old Residency in Munich (1914) by Adolf Hitler. A book called *Adolf Hitler as Painter and Draftsman* was published in 1983 in Switzerland. It catalogs some 750 of Hitler's watercolors, oils, and sketches. Offered to several New York publishing houses, it "was rejected on the grounds that it risked making Hitler appear human."

to be authentic. In 1939, Hitler was visiting the family of Siegfried Wagner (the composer's son) at Bayreuth. The children adored him and called him by a special nickname, "Uncle Wolf." Siegfried's wife, Winifred, was a particular friend, and Hitler said to her of his *Rienzi* epiphany, "That was when it all began" — "it" being the process that turned an unpromising boy into the great führer. Hitler also told the *Rienzi* story to members of his inner circle (such as his architect, Albert Speer) and to the generals on his staff.

Of course, this doesn't mean that if young Adolfus had skipped *Rienzi,* the world could have skipped World War II and the Holocaust. But even historians who are skeptical of

Richard Wagner (1813–1883).

the *Rienzi* story do not deny that Wagner's sprawling hero sagas—with their Germanic gods and knights, their Valkyries
and giants, their stark portrayals of good and evil—helped
shape Hitler's character.

Wagner was not just a brilliant composer. He was also an
extreme German nationalist, a prolific writer of inflamma-
tory political tracts, and a virulent anti-Semite who wrote of
a "grand solution" to the Jewish menace long before the Nazis put one in place. Hitler worshiped Wagner like a god and
called Wagner's music his religion. He attended parts of Wagner's Ring Cycle more than 140 times, and as führer he never
traveled anywhere without his Wagner recordings. He considered the composer to be his mentor, his model, his one true
ancestor. According to André François-Poncet, the French am-
bassador to Berlin in the 1930s, Hitler "'lived' Wagner's work,
he believed himself to be a Wagnerian hero; he was Lohen-

grin, Siegfried, Walther von Stolzing, and especially Parsifal."
He saw himself, in other words, as a modern knight locked in
a struggle with evil.

The acclaimed Hitler biographer Joachim Fest agrees with
François-Poncet: "For the Master of Bayreuth [Wagner] was
not only Hitler's great exemplar, he was also the young man's
ideological mentor . . . [Wagner's] political writings, together
with the operas, form the entire framework for Hitler's ideol-
ogy . . . Here he found the 'granite foundations' for his view
of the world." Hitler himself said that "whoever wants to understand National Socialist Germany must understand the
works of Wagner."

INK PEOPLE

The characters in fiction are just wiggles of ink on paper (or
chemical stains on celluloid). They are ink people. They live
in ink houses inside ink towns. They work at ink jobs. They
have inky problems. They sweat ink and cry ink, and when
they are cut, they bleed ink. And yet ink people press effort-
lessly through the porous membrane separating their inky
world from ours. They move through our flesh-and-blood
world and wield real power in it. As we have seen, this is spectacularly true of sacred fictions. The ink people of scripture
have a real, live presence in our world. They shape our behaviors and our customs, and in so doing, they transform societies and histories.

This is also true of ordinary fiction. In 1835, Edward Bulwer-Lytton wrote a novel called *Rienzi*. The young Richard
Wagner was inspired by the novel and decided to base an opera on it. Bulwer-Lytton conjured people out of paper and
ink. Wagner put those ink people onstage and told their story

James Koehnline's *Literature* (2007).

in song. Those songs changed Adolf Hitler and, through Hitler, the world. Wagner's ink people—Siegfried, Parsifal, Rienzi—may have been significant in the wild mix of factors that brought on the worst war in history, and the worst genocide.

The eleventh edition of *The Encyclopaedia Britannica* claimed vast power for literary art, saying it has had "as much effect upon human destiny" as the taming of fire. But not everyone thinks so. W. H. Auden wrote that "poetry makes nothing happen," and Oscar Wilde wrote that "all art is quite useless." Stories, in this view, are relatively inert in their effects. After all, most people are not stupid. They know the difference between reality and fantasy, and they resist being manipulated.

Until very recently, this debate was driven largely by anecdote. The most famous by far involves the plight of an ink person named Eliza Harris. Young and beautiful, spirited and good, Eliza was a slave who belonged to Arthur Shelby. Rather than see her small son Harry sold "down the river" to the much rougher plantations of the Deep South, Eliza ran for the North. An account of her flight was serially published beginning in 1851 in the newspaper *National Era*. Readers held their breath as Eliza stood on the south bank of the Ohio River, looking out over the churning expanse of ice floes that separated the slave state of Kentucky from the free state of Ohio. At her back, the slave catchers were already in sight, closing fast. Holding little Harry in her arms, Eliza stepped onto the uncertain ice. Then, leaping and slipping from ice floe to wobbly ice floe, she made it to the other side, and eventually to freedom in Canada.

In 1852, the story of Eliza's terrible struggle, and those of another slave from the Shelby estate, Uncle Tom, was republished in book form. It would become, with the exception of the Bible, the best-selling book of the nineteenth century. *Uncle Tom's Cabin* polarized the American public. By showing the cruelty of slavery, the book roused abolitionist sympathies in the North. And by depicting slavery as a hellish institution ruled over by brutes, the book helped galvanize the South in slavery's defense. The book's representative slave owner, Simon Legree, is a sadistic monster who has a fist like a "blacksmith's hammer." He shakes that hammer in the face of his slaves and says, "This yer fist has got as hard as iron knocking down niggers. I never seen the nigger yet I couldn't bring down with one crack."

When, in the midst of the Civil War, President Abraham Lincoln met Harriet Beecher Stowe, he famously said, "So

Eliza Harris crossing the Ohio River. A promotional poster for an 1881 theatrical adaptation of *Uncle Tom's Cabin.*

you're the little woman who wrote the book that made this great war." Lincoln went a little far in his flattery, but historians agree that *Uncle Tom's Cabin* "exerted a momentous impact on American culture (and continues to do so)," inflaming the passions that brought on the most terrible war in American history. Moreover, it affected international opinion in important ways. As the historian Paul Johnson has written, "In Britain, the success of the novel helped to ensure that . . . the British, whose economic interest lay with the South, remained strictly neutral." If the British had jumped into the fight, the outcome may have been different.

People who believe that story systematically shapes individuals and cultures can cite plenty of evidence beyond *Rienzi* and *Uncle Tom's Cabin:* the way D. W. Griffith's 1915 epic film, *The Birth of a Nation,* resurrected the defunct Ku Klux Klan; the way the film *Jaws* (1975) depressed the economies of coastal holiday towns; the way Charles Dickens's *A Christmas Carol* (1843) is—in the words of Christopher Hitchens—responsible for much of "the grisly inheritance that is the modern version of Christmas"; the way *The Iliad* gave Alexander the Great a thirst for immortal glory (the eighteenth-century novelist Samuel Richardson asked, "Would Alexander, madman as he was, have been so *much* a madman, had it not been for Homer?"); the way the publication of Goethe's *The Sorrows of Young Werther* (1774) inspired a spate of copy-cat suicides; the way novels such as *1984* (George Orwell, 1948) and *Darkness at Noon* (Arthur Koestler, 1940) steeled a generation against the nightmare of totalitarianism; the way stories such as *Invisible Man* (Ralph Ellison, 1952), *To Kill a Mockingbird* (Harper Lee, 1960), and *Roots* (Alex Haley, 1976) changed racial attitudes around the world.

The list could go on and on. But it actually proves very little, because the interesting question isn't whether stories sometimes change people or influence history, but whether those changes are predictable and systematic. A skeptic might yawn at this list and say, "Anecdotes don't make a science."

In recent decades, roughly corresponding with the rise of TV, psychology has begun a serious study of story's effects on the human mind. Research results have been consistent and robust: fiction *does* mold our minds. Story—whether delivered through films, books, or video games—teaches us facts about the world; influences our moral logic; and marks us with fears, hopes, and anxieties that alter our behavior, perhaps even our personalities. Research shows that story is constantly nibbling and kneading us, shaping our minds without our knowledge or consent. The more deeply we are cast under story's spell, the more potent its influence.

Most of us believe that we know how to separate fantasy

and reality—that we keep information gathered from fiction safely quarantined from our stores of general knowledge. But studies show that this is not always the case. In the same mental bin, we mix information gleaned from both fiction and nonfiction. In laboratory settings, fiction can mislead people into believing outlandish things: that brushing their teeth is bad for them, that they can "catch" madness during a visit to a mental asylum, or that penicillin has been a disaster for humankind.

Think about it: fiction has probably taught you as much about the world as anything else. What would you actually know about, say, police work without television shows such as *CSI* or *NYPD Blue*? What would I know about tsarist Russia without Tolstoy and Dostoyevsky? Not much. What would I know about British naval life in the Napoleonic era if not for the habit-forming "Master and Commander" novels of Patrick O'Brian? Even less.

And it is not just static information that is passed along through stories. Tolstoy believed that an artist's job is to "infect" his audience with his own ideas and emotions—"the stronger the infection, the better is the art as art." Tolstoy was right—the emotions and ideas in fiction are highly contagious, and people tend to overestimate their immunity to them.

Take fear. Scary stories leave scars. In a 2009 study, the psychologist Joanne Cantor showed that most of us have been traumatized by scary fiction. Seventy-five percent of her research subjects reported intense anxiety, disruptive thoughts, and sleeplessness after viewing a horror film. For a quarter of her subjects, the lingering effects of the experience persisted for more than six years. But here's what's most interesting about Cantor's study: She didn't set out to study movies in particular. She set out to study fear reactions across all mass media—television news, magazine articles, political speeches, and so on. Yet for 91 percent of Cantor's subjects, scary films—not real-world nightmares such as 9/11 or the Rwandan genocide by machete—were the source of their most traumatic memories.

The emotions of fiction are highly contagious, and so are the ideas. As the psychologist Raymond Mar writes, "Researchers have repeatedly found that reader attitudes shift to become more congruent with the ideas expressed in a [fiction] narrative." In fact, fiction seems to be more effective at changing beliefs than nonfiction, which is *designed* to persuade through argument and evidence. For example, if we watch a TV program showing a sexual encounter gone wrong, our own sexual ethics will change. We will be more critical of premarital sex and more judgmental of other people's sexual choices. If, however, the show portrays a positive sexual encounter, our own sexual attitudes will move toward the permissive end of the spectrum. These effects can be demonstrated after a single viewing of a single episode of a prime-time television drama.

As with sex, so too with violence. The effects of violence in the mass media have been the subject of hundreds of studies over the past forty years. This research is controversial, but it seems to show that consuming a lot of violent fiction has consequences. After watching a violent TV program, adults and children behave more aggressively in lab settings. And long-term studies suggest a relationship between the amount of violent fiction consumed in childhood and a person's actual likelihood of behaving violently in the real world. (The opposite relationship also holds: consuming fiction with prosocial themes makes us more cooperative in lab settings.)

It is not only crude attitudes toward sex and violence that

are shaped by fiction. As mentioned in the last chapter, studies have shown that people's deepest moral beliefs and values are modified by the fiction they consume. For example, fictional portrayals of members of different races affect how we view out-groups. After white viewers see a positive portrayal of black family life—say, in *The Cosby Show*—they usually exhibit more positive attitudes toward black people generally. The opposite occurs after white people watch hard-core rap videos.

What is going on here? Why are we putty in a storyteller's hands? One possibility, to borrow the words of Somerset Maugham, is that fiction writers mix the powder (the medicine) of a message with the sugary jam of storytelling. People bolt down the sweet jam of storytelling and don't even notice the undertaste of the powder (whatever message the writer is communicating).

A related explanation comes from the psychologists Melanie Green and Timothy Brock. They argue that entering fictional worlds "radically alters the way information is processed." Green and Brock's research shows that the more absorbed readers are in a story, the more the story changes them. Fiction readers who reported a high level of absorption tended to have their beliefs changed in a more "story-consistent" way than those who were less absorbed. Highly absorbed readers also detected significantly fewer "false notes" in stories—inaccuracies, infelicities—than less transported readers. Importantly, it is not just that highly absorbed readers detected the false notes and didn't care about them (as when we watch a pleasurably idiotic action film); these readers were unable to detect the false notes in the first place.

And in this there is an important lesson about the molding power of story. When we read nonfiction, we read with

Anton Chekhov (1860–1904). Stories change our beliefs and maybe even our personalities. In one study, psychologists gave personality tests to people before and after reading Chekhov's classic short story "The Lady with the Little Dog." In contrast to a control group of nonfiction readers, the fiction readers experienced meaningful changes in their personality profiles directly after reading the story—perhaps because story forces us to enter the minds of characters, softening and confusing our sense of self. The personality changes were "modest" and possibly temporary, but the researchers asked an interesting question: might many little doses of fiction eventually add up to big personality changes?

our shields up. We are critical and skeptical. But when we are absorbed in a story, we drop our intellectual guard. We are moved emotionally, and this seems to leave us defenseless.

There is still a lot to be discovered about the extent and magnitude of story's sculpting power. Most current research is based on extremely low doses of story. People can be made to think differently about sex, race, class, gender, violence, ethics, and just about anything else based on a single short story or television episode.

Now extrapolate. We humans are constantly marinating ourselves in fiction, and all the while it is shaping us, changing us. If the research is correct, fiction is one of the primary sculpting forces of individuals and societies. Anecdotes about those rare ink people, such as Rienzi or Uncle Tom, who vault across the fantasy-reality divide to change history are impressive. But what is more impressive, if harder to see, is the way stories are working on us all the time, reshaping us in the way that flowing water gradually reshapes a rock.

HOLOCAUST, 1933

Adolf Hitler is a potent example of the ways that story can shape individuals and histories, sometimes disastrously. The musical stories that Hitler most loved did not make him a better person. They did not humanize him, soften him, or extend his moral sympathies beyond his own in-group. Quite the opposite. Hitler was able to drive the world into a war that cost sixty million lives not in spite of his love of art but at least partly because of it.

Hitler ruled through art, and he ruled *for* art. In his book *Hitler and the Power of Aesthetics,* Frederic Spotts writes that Hitler's ultimate goals were not military and political; they were broadly artistic. In the new Reich, the arts would be supreme. Spotts criticizes historians who treat Hitler's devotion to the arts as insincere, shallow, or strictly propagandistic. For Spotts, "Hitler's interest in the arts was as intense as his racism; to disregard the one is as profound a distortion as to pass over the other."

On the night of May 10, 1933, Nazis across Germany indulged in an ecstasy of book burning. They burned books written by Jews, modernists, socialists, "art-Bolsheviks," and

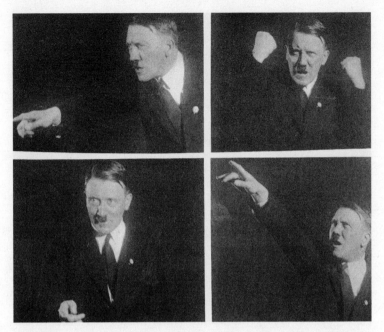

Adolf Hitler practicing theatrical poses for use in speaking performances. Hitler once called himself "the greatest actor in Europe." Frederic Spotts agrees, arguing that Hitler's mastery of public theater helped him mesmerize and mobilize the German people. After watching the Nazi propaganda film *Triumph of the Will* (1935) fifteen times, the singer David Bowie said, "Hitler was one of the first great rock stars. He was no politician. He was a great media artist. How he worked his audience! He made women all hot and sweaty and guys wished they were the ones who were up there. The world will never see anything like that again. He made an entire country a stage show."

writers deemed "un-German in spirit." They were purifying German letters by fire. In Berlin, tens of thousands gathered in the firelight to hear Propaganda Minister Joseph Goebbels shout, "No to decadence and moral corruption! . . . Yes to decency and morality in family and state! I consign to the flames the writings of Heinrich Mann, Bertolt Brecht, Ernst Gläser, Erich Kästner." And with them went the ink children of Jack

London, Theodore Dreiser, Ernest Hemingway, Thomas Mann, and many others.

The Nazis, deeply inspired by Wagner's musical stories, understood that ink people are among the most powerful and dangerous people in the world. And so they committed a holocaust of undesirable ink people so there would be fewer barriers to a holocaust of real people.

Among the books burned that night in 1933 was the play *Almansor* (1821) by the German Jewish writer Heinrich Heine. The play contains this famous and prophetic line: "Where they burn books, they will also ultimately burn people."

Life Stories

"How old was I when you first took me in a boat?"

"Five and you nearly were killed when I brought the fish in too green and he nearly tore the boat to pieces. Can you remember?"

"I can remember the tail slapping and banging and the thwart breaking and the noise of the clubbing. I can remember you throwing me into the bow where the wet coiled lines were and feeling the whole boat shiver and the noise of you clubbing him like chopping a tree down and the sweet blood smell all over me."

"Can you really remember that or did I just tell it to you?"

"I remember everything."

— ERNEST HEMINGWAY, *The Old Man and the Sea*

DAVID WAS, AT THIRTY-ONE, a drunk and a junkie. The day before he got fired was St. Patrick's Day, and David got really tuned up on booze and cocaine. The next day was the worst of David's life. He dragged himself into the offices of the magazine where he worked, feeling cadaverous, and may have brought himself back to life by snorting lines from the bottom of his desk drawer. The editor called David to his office and told him that keeping his job meant going to rehab. David said, "I'm not done yet."

David obediently cleaned out his desk and then hit the bars with his best friend, Donald. They drank whiskey and beer all day, and kept drinking deep into the night. Between drinks and bars, they snorted coke in bathrooms and alleys. Bouncers ejected them from one club for various outrages, and then the two friends tussled with each other in the club parking lot. Donald got sore and went home. David went to a different bar to drink and to seethe about Donald abandoning him.

David rang Donald at home. "I'm coming over," he threatened.

"Don't do it," Donald replied. "I have a gun."

"Oh really? Now I'm coming over for sure."

David may have walked to Donald's house or driven; he can't remember. But when he arrived, he quickly grew bored of knocking at the locked front door and instead attacked it with his feet and his shoulders.

When Donald opened the door, he had the pistol in his hand. He warned David to calm down or he would call the police. David shouldered Donald aside and stumbled for the kitchen, punching through a window on the way. David seized the kitchen phone and presented it to Donald, his hand streaming blood. "All right, call 'em motherfucker! Call 'em! Call the goddamned cops!"

To David's surprise, Donald did. Within minutes, a patrol car pulled up at the house. David fled through the back door, racing toward his apartment eight blocks away, hiding in bushes and alleys as the cops gave chase. David reached his apartment and, bleeding steadily from his hand, passed out.

Twenty years later, David Carr was a columnist for the

New York Times, working on a memoir of his life. Early in the process, he interviewed his old friend Donald. First, David told Donald what he remembered of the worst day of his life. Donald listened, nodding and laughing through most of the story: that was how he remembered it, too. But when David got to the part about the gun, Donald frowned.

Donald said that David's account was right except for one detail: David was the one with the gun in his hand.

Carr writes in his memoir, *The Night of the Gun,* "People remember what they can live with more often than how they lived."

For his memoir, Carr didn't rely solely on his memory. He went out and reported extensively on his own life. He did this for two reasons. First, Carr had spent much of his life drunk and stoned out of his mind; he knew his memory was cooked. Second, he was writing in the turbulent wake of James Frey's fraudulent memoir, *A Million Little Pieces,* and he knew readers would be skeptical of over-the-top details in (yet another) ex-junkie's memoir of triumph over addiction.

In *A Million Little Pieces,* Frey describes his sordid career as a drunk, crack addict, and outlaw who finally got straight. It is a gripping read, and an uplifting one. It was the uplifting part that got Frey on *The Oprah Winfrey Show.* And because Frey got on *Oprah,* he sold truckloads of books and made millions in royalties. It was the gripping part of *A Million Little Pieces* that got Frey on the Smoking Gun as the subject of a masterpiece of debunking called "A Million Little Lies."

Most of the "stranger than fiction" details of Frey's book were actually just regular old fiction. Some points were merely embellished. For example, during his second appearance on *Oprah,* Frey said that the story about having root canals during rehab was absolutely true—except for that part about refusing Novocain. Other details were entirely fabricated, such as the one about Frey being wanted by the law across several states. Pundits howled in protest. Frey crawled back on TV to let Oprah ritually eviscerate him. The rest of us leaned back and enjoyed the spectacle.

But Frey's inventions were nothing compared to some other recent memoirs. In *Memoir: A History*, journalist Ben Yagoda shows that lying memoirs are as old as books, but that the past forty years "will probably be remembered as the golden age of autobiographical fraud. There has been about a scandal a year, and sometimes more than that." For example, there was *Misha: A Mémoire of the Holocaust Years* (1997). *Misha* tells the story of a little Jewish girl's miraculous survival in Nazi Germany. Her adventures included getting trapped in the Warsaw ghetto, stabbing a Nazi rapist, trekking across Europe on foot, and being adopted—like Mowgli in Rudyard Kipling's *The Jungle Book* (1894)—by a pack of kindly wolves. Only none of it was true—not the part about the kindly wolves, not even the part about Misha (real name Monique De Wael) being Jewish.

And then there's my favorite example, *The Blood Runs Like a River Through My Dreams* (2000), one of three highly regarded memoirs by a Native American writer called Nasdijj, who was the victim of fetal alcohol syndrome, homelessness, and rampant white prejudice. Elsewhere, Nasdijj wrote, "My literary lineage is Athabaskan. I hear Changing Woman in my head. I listen to trees, rocks, deserts, crows, and the tongues of the wind. I am Navajo and the European things you relate so closely to often simply seem alien and remote. I do not know them. What I know is the poetry of peyote, the songs of drums, and the dancing of the boy twins, Tobajishinchini and Neyaniinezghanii." And so on.

Nasdijj turned out to be Timothy Barrus, a white North Carolina writer of sadomasochistic gay erotica. Another celebrated Native American memoir, *The Education of Little Tree* (1976), turned out to have been written by a white guy named Asa Carter (pen name Forrest Carter), a former leader of a paramilitary organization called "the Original Ku Klux Klan of the Confederacy."

While these examples are extreme, most memoirs are strewn with unhidden falsehood. Crack open the average

Asa Carter (pen name Forrest Carter) railing against integration in Clinton, Tennessee. Carter was a "virulent segregationist, former Klansman, speech-writer for [Alabama governor] George Wallace and professional racist." His fraudulent memoir of Native American boyhood has sold more than 2.5 million copies.

memoir, and you will find the story of a person's life. The story will be told in a clear story grammar, complete with problem structure and good-guy, bad-guy dynamics. The dramatic arcs are suspiciously familiar; the tales of falling down and rising up, suspiciously formulaic. Amazing things—dramatic, emotional—happen to memoirists with amazing frequency. Memoirists can recall scenes and conversations from their childhoods in unbelievably (that is, "not believable") piquant detail.

Some critics argue that most memoirs, not just the brazenly fraudulent ones, should be shelved in the fiction section of bookstores. Memoirists don't tell true stories; they tell "truthy" ones. Like a film that dramatizes historical events, all memoirs should come with a standard disclaimer: "This book is *based* on a true story."

Every time there is a new memoir scandal, we huff about being tricked. We moan that the writer has betrayed a sacred trust, and we brand the writer as a cheat, a liar, a scoundrel. And then many of us rush out to buy the next grippingly truthy memoir of tribulation and overcoming, of sexual abuse, alcoholism, anal sex enthusiasm (Toni Bentley's *The Surrender*, 2004), or whatever.

But before we stone memoirists for the way they tell their stories, we should look more closely at the way we tell our own. We spend our lives crafting stories that make us the noble—if flawed—protagonists of first-person dramas. A life story is a "personal myth" about who we are deep down—where we come from, how we got this way, and what it all means. Our life stories are who we are. They are our identity. A life story is not, however, an objective account. A life story is a carefully shaped narrative that is replete with strategic forgetting and skillfully spun meanings.

Like any published memoir, our own life stories should also come with a disclaimer: "This story that I tell about myself is only *based* on a true story. I am in large part a figment of my own yearning imagination." And it's a good thing, too. As we will see, a life story is an intensely useful fiction.

"MEMORY, OF COURSE, IS NEVER TRUE"

Scientific history was made in 1889, in the small French town of Nancy, when sixteen-year-old Marie G. reported a terrible crime. She was walking down the hall in her boarding house when she heard sounds of furniture squeaking and banging, along with muffled whimpers, grunts, and moans. Marie stopped outside the room of an old bachelor. She looked to her left and her right, and, seeing no one in the dim hallway, she stooped and pressed her eye to the bright keyhole in the old man's door. The terrible image was seared into her memory: an old man raping a young girl; the girl's wide eyes; the blood; the girl crying out through the gag in her mouth. Wringing her hands, Marie fled down the corridor to her room.

The magistrate listened carefully, but skeptically, to Marie, then told her that he would not be referring the case to the police. Marie was distraught; she told the magistrate that she would stand up in court and swear to her story "before God and man." The magistrate shook his head. He had made up his mind before Marie had even entered the room.

In 1977, the psychologists Roger Brown and James Kulik coined the term "flashbulb memories" to describe photo-perfect recollections of John F. Kennedy's assassination. People vividly remembered where they were, what they were doing, and who they were with when they heard the awful news.

Subsequent research on flashbulb memory has shown that Brown and Kulik were both right and wrong. We really do vividly remember the big and traumatic moments of our lives, but the details of these memories can't be trusted.

For example, the day after the space shuttle *Challenger* exploded in January 1986, researchers asked people how they had heard about the disaster, how they had felt, and what they had been doing. The researchers then returned to the same people two and a half years later and asked them exactly the same questions. As the psychologist Lauren French and her colleagues later put it, "For a quarter of the people, not one detail was consistent between the two reports. On average, fewer than half of the details reported in the follow-up questionnaire matched those reported in the original questionnaire.

A crowd outside a New York City radio shop waits for news on the assassination of President John F. Kennedy in November 1963.

Not one person was completely consistent. What is even more interesting is that two and a half years later, most people were highly confident about the accuracy of their memories."

The signature flashbulb moment of our age is 9/11, which led to a bonanza of false-memory research. The research shows two things: that people are extremely sure of their 9/11 memories and that upward of 70 percent of us misremember key aspects of the attacks. For example, on the morning of September 11, 2001, do you remember when you saw footage of the first plane hitting the World Trade Center? President George W. Bush did. On December 4, 2001, he described learning about the attacks:

> I was in Florida. And my chief of staff, Andy Card—actually I was in a classroom talking about a reading program that works. And I was sitting outside the classroom waiting to go in, and I saw an airplane hit the tower—the TV was obviously on, and I use[d] to fly myself, and I said, "There's one terrible pilot." And I said, "It must have been a horrible accident." But I was whisked off there—I didn't have much time to think about it, and I was sitting in the classroom, and Andy Card, my chief who was sitting over here walked in and said, "A second plane has hit the tower. America's under attack."

For the conspiracy theorists of the so-called 9/11 Truth movement, Bush's statement was a smoking gun. On the morning of the attacks, there was no footage available of the first plane hitting the tower. Therefore, by Truther logic, Bush must have been watching film shot by the government operatives who actually brought the towers down. The headline at FreeWorldAlliance.com screamed, "Bush Slip Reveals Total 9/11 Complicity."

But Bush wasn't alone in falsely remembering seeing the

first plane hit the tower on 9/11. In one study, 73 percent of research subjects confidently misremembered watching, horrified, as the first plane plowed into the North Tower on the morning of September 11.

Similarly, according to the psychologist James Ost, many British people recall seeing nonexistent footage of the Paris car crash that killed Princess Diana, and four out of ten Brits remember seeing terrible images of London's 7/7 bombings that simply don't exist. In short, flashbulb memory research shows that some of the most confident memories in our heads are sheer invention.

Which brings us back to Marie G. When she reported the rape, the skeptical magistrate walked her not to the police station to file charges, but to the psychiatric office of Hippolyte Bernheim, just as Bernheim had asked him to. With the magistrate looking on, Bernheim had Marie lie down on his consulting couch, whereupon Marie repeated the whole grisly tale. Bernheim asked Marie several questions. Are you sure of what you saw? Are you sure you weren't dreaming or hallucinating? When Marie answered yes to these questions, Bernheim was ready with one more: was Marie sure that Bernheim hadn't planted a fake rape memory in her mind?

Marie was Bernheim's patient. She was, in his words, "an intelligent woman," gainfully employed as a shoemaker. Bernheim was supposed to be treating Marie for sleepwalking and nervous symptoms, but he promptly began running experiments on her as well. He wanted to see if he could create a "retroactive hallucination"—a fraudulent memory that Marie could not distinguish from the real events of her life. Bernheim started relatively small. For example, he made Marie believe that, during one of their sessions, she had suffered an acute bout of diarrhea and had to keep rushing off to the bathroom. He made her believe that she had recently bumped her nose so hard that blood gushed from her nostrils like water from a tap.

Satisfied that he could plant relatively mundane memories, Bernheim decided to test himself. In a stroke that may yet earn Bernheim a plaque in the mad scientist hall of fame, the psychiatrist decided to burden Marie with the memory of a hideous child rape. Even afterward, when Bernheim told Marie that the memory was fake, she would not believe him. Marie's memory of the crime was so vivid that it was, for her, an "incontestable reality."

Bernheim used hypnosis to plant memories in Marie's mind. As a result, his claims were met with a lot of skepticism. Memory continued to be seen as a trustworthy system. Memory was fact.

Hippolyte Bernheim (1840–1919).

Then came "the great sex panic of the 1990s." Across the country, psychiatrists, hypnotherapists, and other healers were "recovering" repressed memories of childhood abuse in adult subjects. But many argued that the healers were inadvertently creating false memories, not excavating real ones. For skeptics of recovered memory, the main difference between Bernheim's techniques and those of modern therapists was that Bernheim knew what he was doing, while most modern therapists did not.

The controversy raged, and psychologists set out to settle the issue scientifically, probing the memory system for weaknesses like hackers attacking a computer system. They found that memory was much less trustworthy than anyone had previously suspected.

In a classic experiment, Elizabeth Loftus and her colleagues gathered information from independent sources about undergraduate students' childhoods. The psychologists then brought students into the lab and went over lists of actual events in their lives. The lists were Trojan horses that hid a single lie: When the student was five years old, the psychologists claimed, he wandered away from his parents in a mall. His parents were frightened, and so was he. Eventually an old man reunited him with his parents. At first, the students had no memory of this fictional event. But when they were later called back to the lab and asked about the mall episode, 25 percent of them said they remembered it. These students not only recalled the bare events that the researchers had supplied, but they also added many vivid details of their own.

This study was among the first of many to show how shockingly vulnerable the memory system is to contamination by suggestion. In lab settings, psychologists were able to plant clear childhood memories of meeting Bugs Bunny at Disney-

land (even though Bugs is not a Disney character), of hors-
ing around at a wedding and spilling a whole bowl of punch
on the bride's parents, of taking a ride in a hot-air balloon, of
being hospitalized after being attacked by dogs or other children, and of seeing a cargo plane crash into a Dutch apartment building.

This research is profoundly unsettling. If we can't trust our
memories about the big things in life—9/11, sexual abuse, being hospitalized after a dog attack—how can we trust it about
the small things? How can we believe that anything in our
lives was as we remember it, especially since we are every bit as
confident in our false memories—our "retroactive hallucinations"—as we are in our true ones?

Sure enough, ordinary memory is failing us all the time,
even without the meddling of psychologists. It's not just that
we forget things. It's that what we remember is inaccurate,
sometimes grossly so. For example, in one study men were interviewed right after high school graduation and again decades later. In the first interview, 33 percent of them reported
receiving corporal punishment in high school. When the same
men were interviewed thirty years later, 90 percent said they
had experienced such punishment. In other words, nearly 60
percent of the men had fabricated authentic-seeming memories of being physically beaten by school authorities.

Memory researchers caution that these results shouldn't be
pushed too far. They point out that memory obviously does a
pretty good job of preserving the basic contours of our lives.
My name really is Jonathan Gottschall. I really went to Platts-
burgh High School. I really was raised by Marcia and Jon.
One fateful day in the mid-1980s, I really did sock my little brother, Robert, in the back of the skull as he innocently
fished in the deep freeze for a bean and cheese burrito. (Or

did I? Yes, I did. Robert confirms it. But he thinks he may have been after Toaster Strudel.) Yet the research shows that our memories are not what we think they are. Most of us believe that they are filled with reliable information that we can access whenever we want to. But it's not quite so simple. Like the amnesiac lead character in the 2000 film *Memento*, we all go through life tattooed with indelible memories that didn't happen the way we remember them.

That childhood memory of crashing your new bike on your birthday may get blended together with memories of other accidents and other birthdays. When we recall something from the past, we don't access a file that says "Bike Accident, Eight Years Old." Pieces of that memory are scattered through the brain. Memories for sight, sound, taste, and smell are stored in different locations. When we recall the bike accident, we don't queue up a videotape; we recall bits of data from all around the brain. This data is then sent forward to the storytelling mind—our little storytelling Holmes—who stitches and pastes the scraps and fragments into a coherent and plausible re-creation of what might have occurred, taking his usual poetic license.

Put differently, the past, like the future, does not really exist. They are both fantasies created in our minds. The future is a probabilistic simulation we run in our heads in order to help shape the world we want to live in. The past, unlike the future, has actually happened. But the past, as represented in our minds, is a mental simulation, too. Our memories are not precise records of what actually happened. They are reconstructions of what happened, and many of the details—small and large—are unreliable.

Memory isn't an outright fiction; it is merely a fictionalization.

HEROES OF OUR OWN EPICS

In view of memory's frailties, omissions, and inventions, some researchers have concluded that it just doesn't work very well. But, as the psychologist Jerome Bruner observes, memory may "serve many masters aside from truth." If the purpose of memory is to provide a photo-perfect record of the past, then memory is deeply flawed. But if the purpose of memory is to allow us to live better lives, then the plasticity of memory may actually be useful. Memory may be faulty by design.

As the psychologists Carol Tavris and Eliot Aronson put it, memory is an "unreliable, self-serving historian . . . Memories are often pruned and shaped by an ego-enhancing bias that blurs the edges of past events, softens culpability, and distorts what really happened." Put differently, we misremember the past in a way that allows us to maintain protagonist status in the stories of our own lives.

Even truly awful people usually don't know that they are antagonists. Hitler, for example, thought he was a brave knight who would vanquish evil and bring on a thousand years of paradise on earth. What Stephen King wrote about the villain in his novel *Misery* applies to real villains as well: "Annie Wilkes, the nurse who holds Paul Sheldon prisoner in *Misery*, may seem psychopathic to us, but it's important to remember that she seems perfectly sane and reasonable to herself—heroic, in fact, a beleaguered woman trying to survive in a hostile world filled with cockadoodie brats." Studies show that when ordinary people do something wrong—break a promise, commit a murder—they usually fold it into a narrative that denies or at least diminishes their guilt. This self-exculpatory tendency is so powerful in human life that Steven Pinker calls it the "Great Hypocrisy."

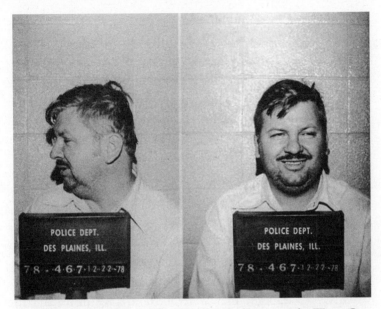

After raping and murdering thirty-three boys in the 1970s, John Wayne Gacy said, "I see myself more as a victim than as a perpetrator . . . I was cheated out of my childhood." He complained that the news media were treating him like a bad guy—like "an asshole and a scapegoat."

This need to see ourselves as the striving heroes of our own epics warps our sense of self. After all, it's not easy to be a plausible protagonist. Fiction protagonists tend to be young, attractive, smart, and brave—all of the things that most of us aren't. Fiction protagonists usually live interesting lives that are marked by intense conflict and drama. We don't. Average Americans work retail or cubicle jobs and spend their nights watching protagonists do interesting things on television, while they eat pork rinds dipped in Miracle Whip.

But on some level, we want to be more like the heroes of fiction, and this means deluding ourselves about who we are and how we got this way. Have you ever seen a photograph of yourself and been rocked by the gap between how you think you look and how fat or saggy or wrinkly or bony you really are? Many people can't understand why they seem to look less attractive in photographs than they do in mirrors. This may be partly about photographic distortion, but it is mainly about the way we unconsciously pose in mirrors — lifting our jaws to stretch away extra chins, raising our eyebrows to smooth away wrinkles and bags — until our best selves appear. We arrange ourselves in the mirror until it tells a flattering lie. This is a good metaphor for what we are doing all the time: building a self-image that improves on the real deal.

When ordinary people are asked to describe themselves, they list many positive qualities and few, if any, negative ones. For example, Thomas Gilovich's book *How We Know What Isn't So* reports that of one million high school seniors surveyed, "70% thought they were above average in leadership ability, and only 2% thought they were below average. In terms of ability to get along with others, nearly *all* students thought they were above average, 60% thought they were in the top 10%, and 25% thought they were in the top 1%!" These self-assessments are obviously wildly out of step with the facts:

it is impossible for a quarter of students to squeeze into the top 1 percent.

We can't blame such outrageous overestimations on the arrogance of youth. We all do it. For example, 90 percent of us think we are above-average drivers, and 94 percent of university professors think they are better than average at their jobs. (I'm honestly surprised that the figure is so low.) College students generally believe that they are more likely than their peers to graduate at the top of their class, earn a big salary, enjoy their work, win awards, and spawn gifted children. College students also believe that they are less likely than others to get fired, get divorced, behave unethically, have cancer, suffer from depression, or have a heart attack.

Psychologists call this "the Lake Woebegone effect": we think we are above average when it comes to just about any positive quality—even immunity to the Lake Woebegone effect. Most of us think we are clear-eyed in our self-assessments. The Lake Woebegone effect applies to other people, not to us. (Be honest, have you been thinking this yourself?) Even when we are willing to admit our failings, we are also likely to denigrate the category. Clumsy at sports? No matter. Sports aren't important. So what's important? The things we are good at. Even though most of us will happily concede that we are not geniuses with movie-star good looks, very few of us will concede that we are actually below average in smarts, social ability, or attractiveness (though half of us are).

It's not that we are all-around optimists; we describe ourselves in much more positive terms than other people, even our friends. We are protagonists, and everyone else is a bit player in our personal drama. And as impressive as we were when we were young, we almost all see ourselves improving with age. For these reasons, the psychologist Cordelia Fine calls the idea of self-knowledge a "farce" and an "agreeable fiction."

Self-aggrandizement starts early and with a vengeance. Small children have grandiose views of their own stellar qualities. This was brought home to me one summer when my daughter Annabel was three. She was convinced that she was breathtakingly fast. How fast? Faster than her father, and certainly faster than her older sister.

The three of us frequently raced from the garden at one corner of the backyard to the playhouse at the other. Annabel always came in second, surging past me as I pretended to falter near the finish line. But Abby, six years old and long-legged, never took a dive. Abby always dusted her tiny sister by furlongs. And yet no matter how many defeats Annabel suffered, they never shook her confidence in her own blazing speed.

After perhaps her tenth loss of the summer, I asked Annabel, "Who's faster, you or Abby?" Her answer was just as proud and confident as it had been after every other crushing defeat: "I'm faster!" Then I asked, "Annabel, who is faster, you or a cheetah?" Annabel knew from watching Animal Planet that cheetahs are scary fast. But she knew that she was scary fast, too. Her answer was a bit less confident: "Me?"

It's different for depressed people. Depressed people have lost their positive illusions; they rate their personal qualities much more plausibly than average. They are able to see, with terrible clarity, that they are not all that special. According to the psychologist Shelley Taylor, a healthy mind tells itself flattering lies. And if it does not lie to itself, it is not healthy. Why? Because, as the philosopher William Hirstein puts it, positive illusions keep us from yielding to despair:

The truth is depressing. We are going to die, most likely after illness; all our friends will likewise die; we are tiny insignificant dots on a tiny planet. Perhaps with the advent of broad intelligence and foresight comes the need for . . . self-deception to keep depression and its consequent lethargy at bay. There needs to be a basic denial of our finitude and insignificance in the larger scene. It takes a certain amount of chutzpah just to get out of bed in the morning.

It is interesting to note that even in this age of Prozac and Zoloft, one of the most common ways of dealing with depression is by talking with a psychotherapist. According to the psychologist Michele Crossley, depression frequently stems from an "incoherent story," an "inadequate narrative account of oneself," or "a life story gone awry." Psychotherapy helps unhappy people set their life stories straight; it literally gives them a story they can live with. And it works. According to a recent review article in *American Psychologist,* controlled scientific studies show that the talking cure works as well as (and perhaps much better than) newer therapies such as antidepressant drugs or cognitive-behavioral therapy. A psychotherapist can therefore be seen as a kind of script doctor who helps patients revise their life stories so that they can play the role of protagonists again—suffering and flawed protagonists, to be sure, but protagonists who are moving toward the light.

All of this research shows that we are the great masterworks of our own storytelling minds—figments of our own imaginations. We think of ourselves as very stable and real. But our memories constrain our self-creation less than we think, and they are constantly being distorted by our hopes and dreams.

Until the day we die, we are living the story of our lives. And, like a novel in process, our life stories are always changing and evolving, being edited, rewritten, and embellished by an unreliable narrator. We are, in large part, our personal stories. And those stories are more truthy than true.

The Future of Story

HUMANS ARE CREATURES of Neverland. Neverland is our evolutionary niche, our special habitat. We are attracted to Neverland because, on the whole, it is good for us. It nourishes our imaginations; it reinforces moral behavior; it gives us safe worlds to practice inside. Story is the glue of human social life—defining groups and holding them together. We live in Neverland because we can't *not* live in Neverland. Neverland is our nature. We are the storytelling animal.

People dream and fantasize, our children romp and dramatize, as much as ever. We are hard-wired to do so. And yet many worry that fiction may be losing its central place in our lives, that we, as a culture, might be leaving Neverland behind. The novel is a young genre, but for a century critics have been writing and rewriting its obituary. If technological changes don't spell its doom, cultural ADHD does. Live theater and poetry are even worse off. Theaters increasingly struggle to make ends meet, and poets trade accusations about who killed poetry. University literature departments are in big trouble, too. English departments have been hemorrhaging majors for decades, and the whole field is in the midst of a great and possibly permanent depression, where two-thirds of Ph.D.'s never find full-time, tenure-track work.

It's not just the "higher" fictional forms that people worry about. "Lower" forms are struggling, too. Many lament the way that cheap and seamy "reality" shows are displacing scripted television. Video games—and other digital entertainments—are also on the rise, drawing audiences away from traditional story. The gaming industry is now much bigger than the book industry, bigger even than the film industry. The 2010 release of Call of Duty: Modern Warfare 3 made more money ($360 million) in the first twenty-four hours than *Avatar* did.

Don't these trends show that fiction is dying a slow death? David Shields thinks so. In his bracing manifesto *Reality Hunger,* Shields proclaims that all forms of conventional fiction are used up, punched out, and withering away. Shields, a former novelist, has tired of his former love, and he wants to help speed this process along: "I come to . . . dispraise fiction, which has never seemed less central to the culture's sense of itself."

Shields oversells his case. Take the novel. Rumors of its demise are exaggerated to the point of absurdity. For some reason, literary intellectuals love to wallow masochistically in the notion that we are living in the last days of the novel. Yet tens of thousands of new novels are published around the world every year, with the total numbers trending up, not down. In the United States alone, a new novel is published every hour. Some of these novels sell by the ton and extend their cultural reach by being turned into films.

When have novels ever delighted more juveniles and adults than Stephanie Meyer's Twilight saga or J. K. Rowling's Harry Potter books (which are twice as long in their entirety as *War and Peace*)? When was the last time any novels made a bigger cultural splash than the postapocalyptic Left Behind series by Tim LaHaye and Jerry Jenkins, with 65 million copies sold? When did authors sell more books to a more devoted public than John Grisham, Dan Brown, Tom Clancy, Nora Roberts, Stephen King, or Stieg Larsson? When has a literary genre outstripped the popularity of the romance novel, which does a cool billion dollars in sales per year? When has any novelist been able to brag—as J. K. Rowling can—of having the same sum in her bank account?

Literary novels are having a harder time of it, but when did they *not* have a harder time? Novelists who target highbrow readers shouldn't complain when those are the only readers they get. Still, over the past decade, many literary novels have found large readerships, including works such as Ian McEwan's *Atonement,* Yann Martel's *Life of Pi,* Jhumpa Lahiri's *The Namesake,* Khaled Hosseini's *The Kite Runner,* Jona-

After queuing until midnight, readers swarm through a California bookstore to snap up copies of *Harry Potter and the Deathly Hallows* (2007).

than Franzen's *Freedom* (which put him on the cover of *Time* magazine), and Cormac McCarthy's *The Road*—which is, I think, about as good a story as a human being can hope to tell.

So whenever you hear that the novel is dead, translate as follows: "I don't like all of those hot-selling novels that are filling up the bestseller lists—so they don't count."

But what if the novel were actually to die or just dwindle into true cultural irrelevance? Would that signal the decline of story? For a bookman like me, the end of the novel would be a very sad thing. But, as David Shields himself stresses, it would not be the end of *story*. The novel is not an eternal literary form. While the novel has ancient precursors, it rose as a dominating force only in the eighteenth century. We were creatures of story before we had novels, and we will be creatures of story if sawed-off attention spans or technological advances ever render the novel obsolete. Story evolves. Like a biological organism, it continuously adapts itself to the demands of its environment.

How about poems? I have a friend, Andrew, who is a talented poet. We sometimes meet for beers, and he laments the decline of poetry's status in the modern world. Poets, he tells me, used to be rock stars. Byron couldn't turn around—in a bar, in a park, in his own drawing room—without getting a pair of silky knickers pegged at him. But, I remind him, poets are still stars; they still get underwear thrown at them. People still love the small, intense stories poets tell. In fact, they love them more than ever—as long as the little stories are accompanied by melody, musical instruments, and the emotion of a singer's voice.

Ours is not the age when poetry died; it is the age when

The 2010 release of *The Anthology of Rap*, Yale University Press's nine-hundred-page collection of rap lyrics, shows that scholars are starting to take hip hop seriously as poetic art. In *Book of Rhymes: The Poetics of Hip Hop*, English professor Adam Bradley argues that rap music is the "most widely disseminated poetry in the history of the world . . . The best MCs—like Rakim, Jay-Z [pictured here], Tupac, and many others—deserve consideration alongside the giants of American poetry."

poetry triumphed in the form of song. It is the age of *American Idol*. It is the age when people carry around ten or twenty thousand of their favorite poems stored on little white rectangles tucked into their hip pockets. It is an age when most of us know hundreds of these poems by heart.

My daughter Abby is dancing in the living room with a wooden spoon in her hand—tossing her hair, swinging her hips, and singing along with her new Taylor Swift CD. When the song reaches an unfamiliar stretch, Abby goes still. She tilts her ear to the speaker and concentrates on learning the words to a story about a modern Romeo and Juliet. Her little sister is dancing wildly in her princess gown and tiara, moving her lips in front of her own wooden spoon and pretending that she, too, knows the words.

My girls live in a particular place and in a particular time. But what is happening in my living room is ancient. As long as there are humans, they will delight in the beat, the melody, and the stories of song.

Along with the fear of things dying, there's a fear of things rising. Video games are a prime example. But do they represent a movement away from story or just a stage in story's evolution? Video games have changed massively from the first arcade games I remember playing as a boy: Asteroids, Pac-Man, Space Invaders. Most hit video games are now intensely story-centric. The gamer controls a virtual character—an avatar (or "mini-me")—who moves through a rich digital Neverland. Pick up a copy of *PC Gamer* magazine, and you will find that—with the exception of sports simulators—most video games are organized around the familiar grammar of problem structure and poetic justice. Marketed mainly to testosterone-drunk young males, the games are usually narratives of lurid but heroic violence. Such games don't take their players out of story; they immerse them in a fantasy world where they get to *be* the rock-jawed hero of an action film.

The plot of the average video game—like that of the average action film—is usually a thin gruel (a guy, a gun, a girl). But we are on the cusp of something richer. As the novelist and critic Tom Bissell notes in his book *Extra Lives,* we are living through the birth of a new form of storytelling where the conventions are still being discovered and refined. Ambitious designers are trying to fuse the appeal of gaming with all the power of musical, visual, and narrative art. For example, the writer/director of Sony PlayStation's Heavy Rain, David

Cage, sought to push game design forward in the revolutionary fashion of *Citizen Kane* (1941). Heavy Rain was conceived not as a video game but as an "interactive film," where you play the roles of several characters who are trying to save a boy from a serial murderer known as the Origami Killer. Throughout Heavy Rain, the "player" inhabits several different characters (including the Origami Killer) and makes decisions that determine how the story will end.

TRUE LIES

The way we experience story on television does seem to be changing, but television is still, in the main, a story-delivery technology. The rise of reality programming, and the displacement of scripted shows, has been greeted as a gruesome sign of the end of fiction, if not of civilization. But tawdry reality shows have risen alongside a true golden age of televised drama (think *The Wire, Mad Men, Breaking Bad, The Sopranos*), and in any case, reality shows are hardly nonfiction. Reality show producers trap clashing strangers in houses or on deserted islands in order to instigate as much dramatic conflict as possible. The "characters" are, to one degree or another, acting. They know that the cameras are on. They know that they are expected to play the part of the hothead drunk or the ingénue or the sexpot, and they know that outrages equal screen time.

Together with teams of editors, reality show writers (yes, writers) take raw footage and twist it into classic story lines. *Extreme Makeover, Queer Eye, The Osbournes, The Real Housewives of New Jersey, Whale Wars, Jon and Kate Plus 8*—these and scores of other reality shows have hewn closely to the universal grammar of storytelling.

For example, on Spike TV's *The Ultimate Fighter*, a group of young men move into a big, fancy house, where they are given all the free booze they can drink but are prohibited from watching TV, reading books, using their phones, or seeing their girlfriends or wives. The point is to make the men as tense and quarrelsome as possible. Like *Survivor*, *The Ultimate Fighter* is a competition that will leave only one person standing. The original twist of *The Ultimate Fighter* is that all of the contestants do one thing especially well: enter an octagonal steel cage and beat other young men into submission. On *The Ultimate Fighter*, you don't get voted off the island; you get pounded out in the cage.

In the show's tenth season, a cage fighter called Meathead defeated a fighter named Scott Junk, seriously injuring Junk's eye. This enraged Junk's friend Big Baby, who immediately rushed into the gym to confront Meathead. Big Baby, an enormous ex–NFL lineman, towered over Meathead with his heavy fists quaking at his sides. He barked hoarsely into Meathead's face, "Swing at me, please! Hit me, bitch! Hit me! Give me a fuckin' reason and I'll kill you, motherfucker!" Meathead stared back at Big Baby, and then he took a small step backward, as if he hoped no one would notice.

On *The Ultimate Fighter*, Big Baby was mainly presented as a good guy and Meathead as a villain. Despite his intimidating build, Big Baby was soft-spoken and kind, with the good manners of a well-raised southern boy. Meathead was the show's outcast. The other fighters thought he lied a lot, and they insulted his intelligence and courage. But both characters were—to use E. M. Forster's term—round, not flat. When Meathead was alone, talking to the camera, he came off as the most cerebral contestant on the show. He seemed to really get that his job was dangerous and that every time he entered the cage, big and scary men were trying their utmost to beat or choke him unconscious.

And while Big Baby seemed like a gentle giant out of central casting, his dark side made him interesting. He reeled back and forth between his two basic personality states: cheerful and enraged. His coach, Rampage Jackson, commented that Big Baby was "the nicest guy in the world . . . who will kill you."

The Ultimate Fighter is based on footage of real people—not actors—negotiating extreme conflict situations. Maybe the show isn't quite fiction, but it also isn't nonfiction. It gives us everything we gravitate toward in stories: extreme and often violent conflict, classic arcs of story and character. But it gives us one thing more: a sense of compelling realism. Most fiction has to strive hard for authenticity. Achieving verisimilitude is a large part of the craft of fiction. Reality programming doesn't have to strive. Robert De Niro's portrayal of a half-mad fighter in *Raging Bull* (1980) is among the greatest performances in cinema history. But it is still a performance—an act of fakery. When De Niro pretends to be crazy with rage, it is not as convincing or terrifying as when Big Baby loses control for real.

At the opposite end of the reality show spectrum is ABC's *SuperNanny*. Each episode begins with the plucky British nanny (Jo Frost) arriving at a home in chaos. The parents are ineffectual. The kids are little monsters. The nanny spends a day or two observing how incredibly screwed up the family is, shaking her head and rolling her eyes for the camera. And then she lays down the law. From the chaos of the household, the nanny forges order—a clean house, wise and loving parents, respectful and well-mannered children. Then she drives off in her frumpy British nanny car, leaving the little family to live happily ever after.

What a fantasy! Few programs match *SuperNanny* for the brazen way it dresses fiction in the robes of "reality." Shows like *SuperNanny* are much less truthful than the average work of fiction. Good fiction tells intensely truthful lies. *SuperNanny* is full of lies, but not truthful ones.

The Ultimate Fighter and *SuperNanny* illustrate that reality shows are not nonfiction. They are just a new kind of fiction, in which the lies and distortions happen mainly in the editing room, not the writing room.

These are undeniably nervous times for people who make a living through story. The publishing, film, and television businesses are going through a period of painful change. But the *essence* of story is not changing. The technology of storytelling has evolved from oral tales, to clay tablets, to hand-lettered manuscripts, to printed books, to movies, televisions, Kindles, and iPhones. This wreaks havoc on business models, but it doesn't fundamentally change story. Fiction is as it was and ever will be:

Character + Predicament + Attempted Extrication

Futurology is a fool's game, but I think the worry that story is being squeezed out of human life is exactly the wrong one. The future will see an intensification, even a perfection, of what draws us to fiction in the first place. The gravitational pull of story is going to increase manyfold. We will be marooned in cyber-Neverlands, and we will like it that way. As one online gamer put it, "The future looks bleak for reality."

BACK TO NEVERLAND

It was a sunny fall day. Ethan and his friends were running through the woods of Indian Springs State Park in Flovilla, Georgia. They dodged between campsites and into the forest, ignoring stares from park rangers and fellow campers. They were hunting monsters, and monsters were hunting them. When they slashed down pretend orcs, they cried, "Die, foul beast!"

In quieter moments, they stayed in character. The hotheaded Magnus Tigersblood apologized to Sir Talon: "I am not wise. I know how to fight and how to draw a few runes." Sir Talon was magnanimous: "Sir, your words are from your heart." All of the heroes hailed from exotic home worlds: the Enchanted Glade, the Empire of Perfect Unity, the Rock of Storms, Goblin City.

Ethan achieved a medieval look by wearing women's clothes purchased at a thrift store—a puffy white blouse and black tights. A fairy named Erin was wearing a satiny dress, ballet shoes, and wings held on by elastic bands. Their weapons were wood or foam wrapped in duct tape. It didn't matter that the costumes were crude or that the blue tarp hanging between trees looked nothing like the entrance to a dungeon. In the players' imaginations, a foam pool toy became a terrifying club, a smudge of dirt transformed a human face into that of a night-stalking goblin, and a cheap piece of costume jewelry became as precious as the grail.

That weekend Ethan and his comrades—Wolf, Aerie, Heinrich Irongear, Dusk Whisper, and all the rest—fought and fled for their lives. They slew terror beaks. They fought "rat-wolf things" in a "cave of extreme foulness." They solved riddles. They cast spells. They fought among themselves and made up.

And in the end, they finally found the Mandrakes, lurking just off a forest path. Mandrakes are half-human, half-vegetable, and all evil. Many a brave warrior and beautiful maiden has stumbled into a thicket of Mandrakes, only to be devoured in seconds.

Here is what it sounded like when Ethan, wielding his foam mace, waded into the Mandrakes, with his fellow heroes battling at his sides:

> Power strike!
> Bam!
> Parry!
> Get in there! Flank him!
> Two more mandrakes!
> Fwaaapppp!
> Power strike 2!
> Dodge!
> Fffff-bapppp-pah-pah-pah!
> Mortal blow!
> Arrrrrgggghhhh!

After making the forest safe for everything good and pure, the heroes returned to their cabins. They called one another by their real names and talked about their spouses and children. And then they climbed into their cars and drove home not to Goblin City or the Rock of Storms, but to the suburbs of Atlanta. Ethan Gilsdorf, fortysomething, said goodbye to his new friends and boarded a plane for Boston, where he was working on a book about the fantasy gaming subculture.

Gilsdorf had just experienced the LARP (live action role

The Russian LARP Stalker is set in the radioactive exclusion zone around the Chernobyl nuclear reactor after a second fictional disaster. The larpers (shown here) must band together to fight off mutant creatures and other dangers.

playing game) called Forest of Doors. In LARP, grownups let their inner children out. They create fantasy scenarios ranging from typical sword and sorcery stuff to sci-fi and secret-agent games. They each develop a rich character, complete with backstory—a heartbroken sorcerer, a prim fairy with a mean streak, a femme fatale with a secret—and then the larpers pretend, sometimes staying in character for days at a time.

LARP is not really a game. It is improv theater without an audience. LARP is grown-up make-believe.

LARP evolved in the 1980s out of tabletop role-playing games (RPGs) such as Dungeons and Dragons, which brought friends together for bouts of cooperative storytelling. RPGs invite us to enter richly conceived fictional worlds not as passive imaginers (as in traditional fiction), but as active characters. RPGs are crossbreeds of games and stories. But to me, the story aspect dominates. "Game" is the name we give to our interactive relationship with the story world.

The stereotypical Dungeons and Dragons player is a pimply, introverted boy who isn't cool and can't play sports or attract girls. From my years of playing Dungeons and Dragons as a kid, and of hanging out with guys who kept playing into adulthood, this stereotype strikes me as pretty accurate. But larpers are another breed altogether. They are the kind of committed übernerds that even Dungeons and Dragons nerds get to snigger at. But no one should be sniggering. Why is acting out Tolkienesque stories considered dorky, when most of us love to sit like lumps in a theater watching actors do the same? Why is LARP considered pathetic, when we practically worship movie stars who prance around in Neverland, hollering and smooching and stabbing and emoting? Larpers are just an extreme example of the Peter Pan principle: humans are the species that won't grow up. We may leave our nurseries behind, but not Neverland.

And there's another reason we shouldn't mock these gamers. RPGs such as Forest of Doors and Dungeons and Dragons point the way to the future of story.

O BRAVE NEW WORLD!

I don't think traditional fiction is dying, and I don't think the universal grammar will ever change. But I do think storytelling will evolve in new directions over the next fifty years. Interactive fiction, in the form of RPGs, will move from the geek fringe to the mainstream. More and more of us will be running around like larpers in la-la land, dreaming up characters and acting them out. But we will be doing so in cyberspace, not in the real world.

Two of the most compelling sci-fi visions of story's future come from Aldous Huxley's *Brave New World* (1932) and *Star Trek: The Next Generation* (1987–1994). In Huxley's dystopian novel, fiction is essentially dead. People flock instead to the "feelies." Feelies are, superficially, a lot like movies, but there are two big differences. First, in a feely you actually *feel* what the characters do. When two people have sex on a bearskin rug, you sense every hair on the rug, your lips mashing with the kisses. Second, a feely isn't really a *story*-delivery technology. It is a *sensation*-delivery technology. Feelies do not explore the human plight. They have zero intellectual content. They are just thrill and shiver. Feelies let people watch their porn and feel it, too.

If the feelies are ever invented, people will, of course, throng to them. But I don't think this would spell the end of story. I think people would want feelies *and* stories. The citizens of Huxley's dystopia are satisfied with feelies, but they are different from us. They have been genetically engineered and culturally conditioned to the point that they are no longer fully human. Story will not go away until we really do cross over into a brave new world—a world in which human nature and nurture are fundamentally changed. Huxley himself seemed to understand this. His novel features only one fully authentic human, John the Savage, who is a deviant partly because he prefers Shakespeare to feelies.

I think the future of fiction will be closer to *Star Trek*'s holodeck than Huxley's feelies. In the fictional universe of *Star Trek: The Next Generation,* the holodeck is capable of authentically simulating just about anything. A holonovel is a fictional work that you enter into, as a character, on the holodeck; it is a technologically sophisticated version of LARP. Like a feely, a holonovel tricks the mind into thinking the story is actually happening. But unlike a feely, the holonovel gives you all the thrill and shiver without stripping away the story.

On the holodeck, Captain Kathryn Janeway enjoys Jane Austen–like holonovels, where she plays the role of a smart, spunky, and much-desired heroine. By contrast, Captain Jean-Luc Picard enjoys solving mysteries as a Raymond Chandler–esque detective named Dixon Hill. *Star Trek*'s holonovel perfects much of what draws us to fiction in the first place: a sense of identification with characters that is complete because we *are* the characters, and a perfect illusion of transportation into an alternate universe.

We may never achieve the technological sophistication that we see in *Star Trek*. But I believe that we are moving in that direction with a specific type of video game called a MMORPG, or massively multiplayer online role-playing game. (Most people pronounce this acronym "Mor-Peg.") In MMORPGs, players become characters in an unfolding story. They move through a physically vast and culturally rich virtual world that they share with thousands of other players. The virtual worlds have their own laws and customs. They have their own linguistic dialects, with vocabularies that can be very difficult for the uninitiated (called "noobs") to master. (Some verbs: to gank, to grief, to nerf, to buff, to debuff, to twink, to gimp, to pwn.) They have warring tribes and thriving economies, with trade amounting to hundreds of millions of real-world dollars per year. Authentic cultures spontaneously develop in MMORPG worlds, and anthropologists write ethnographies about them.

When you enter a MMORPG, you not only enter a distinct physical and cultural space; you also enter a story space. In fact, many MMORPGs are based on popular stories, such

as *The Lord of the Rings, Star Trek,* and *Star Wars.* MMOR-
PGs invite us to become characters in classic hero stories. As
one player put it, playing a MMORPG is like living "inside a
novel as it is being written." Another said, "I'm living inside a
medieval saga. I'm one of the characters in the novel, and, at
the same time, I'm one of the authors."

Take, for example, Blizzard Entertainment's MMORPG,
World of Warcraft (WoW). It is hard to describe WoW in
this compact space, for the same reason it would be difficult
to sum up the physical and cultural concept of Nicaragua or
Norway in a few paragraphs. The ambition of WoW is startling. Its developers aren't game makers; they are world makers. (They tellingly refer to their creation not as a role-playing
game, but as a role-playing *experience.*) WoW's designers are
geek gods carving a virtual world out of the void.

WoW is an online universe made up of many separate
planets, races, factions, cultures, religions, and mutually in-
comprehensible languages. Twelve million real people adventure there (meaning that the population of WoW world
exceeds Nicaragua's and Norway's put together). The sociologist William Sims Bainbridge, who spent two years in WoW
world doing participant-observer research, isn't exaggerating
when he writes that the WoW experience is based on a "tapestry of myths as complex as any in the ancient sagas." There
are books lying around WoW world, such as *The New Horde*
and *Civil War in the Plaguelands,* which your character can
read to learn the lore of the realm. There is a series of novels (stretching to fifteen at this writing) that flesh out WoW's
backstory, develop the major characters, and constrain the
continuous evolution of the online experience. When you enter WoW world, you become a character in an evolving epic
that stretches back to the beginning of time—the first gods,

Screenshot from World of Warcraft showing a female blood elf.

the birth of worlds, and ten thousand years of history charting the rise and fall of races and civilizations.

WoW achieves what it does because it bundles the creativity of many hundreds of collaborators: programmers, writers, social scientists, historians, visual artists, musicians, and others. Most great art is created by individuals, but WoW is the product of hundreds of creative people weaving the power of story art together with visual and sound art. WoW is an art bonanza, and these are still its early days. What will universes such as WoW be like in twenty years? In fifty?

EXODUS

In his book *Exodus to the Virtual World,* economist Edward Castronova argues that we have begun the greatest mass migration in the history of humanity. People are moving en masse from the real to the virtual world. Bodies will always be marooned here on earth, but human attention is gradually "draining" into the virtual world. Tens of millions of MMORPG devotees already spend an average of twenty to thirty hours per week absorbed in online adventures. According to a survey of thirty thousand MMORPG players, about half of all serious players form their most satisfying friendships in-game, and 20 percent consider MMORPG land to be their "true home," while Earth is "merely a place they visit from time to time." The pace of the exodus will increase as technological advances make virtual worlds more and more appealing.

According to Castronova, the exodus will be fueled not only by the attractive force of new virtual worlds—by the strong suction of interactive story—but by the repellent force of real life. Castronova asks us to imagine an average guy named Bob. Bob works in retail—shelving product, sweeping floors, manning the register. He drives through a bleak concrete landscape of big-box stores and fast-food joints. When he bowls, he bowls alone. He is not involved in civic life. He is in no real sense a member of a community, and his life is meaningless. His job asks so little of him, and he produces nothing of lasting value.

But after work, Bob goes online and finds everything that is missing from his life. In MMORPG land, Bob has friends; he may even have a wife. He doesn't live to sell and consume trash; he lives to crusade against evil. In MMORPG land, Bob has big muscles, big weapons, and dangerous magic. He is an essential and respected member of a tight-knit community.

Commentators frequently blame MMORPGs for an increasing sense of isolation in modern life. But virtual worlds are less a cause of that isolation than a response to it. Virtual worlds give back what has been scooped out of modern life.

The virtual world is in important ways more authentically human than the real world. It gives us back community, a feeling of competence, and a sense of being an important person whom people depend on.

Above all, MMORPG worlds are profoundly *meaningful.* As game designer David Rickey put it, people enter MMORPGs to take a daily vacation from the pointlessness of their actual lives. A MMORPG is an intensely meaning-rich environment—a world that seems, in many ways, more worthy of our lives and our deaths. MMORPGs accomplish this, above all, by resurrecting myths. In the virtual world, the myths retain all their power, and the gods are alive and potent. Here is how Warhammer Online describes the sinister warlord Tchar'zanek: "In the lands of the far north, where tribes of savage barbarians worship the abhorrent gods of Chaos, a new champion has risen. His name is heard on the howling of the icy winds and the shrill cries of ravens. It is proclaimed in peals of thunder and whispered in the nightmares of men. He is Tchar'zanek, Chosen of Tzeentch [a god of Chaos], and he will shake the very foundations of the Old World."

So people will increasingly enter MMORPG worlds not only for their positive virtues but also to escape the bleakness of modern life—the feeling that, as game designer Jane McGonigal puts it in the title of her recent book, reality is broken. You might say, "Yeah, but those role-playing geeks all have one thing in common: they are pathetic losers. I'm not. The world of dorks and orcs has nothing to do with me."

True, MMORPGs are not for everybody. But they are still in their infancy. In the decades to come, computing capacity will grow exponentially, and we will move closer and closer to the holonovel. When this happens, story land will outstrip real life in many ways. Many people—especially people like Bob—have already decided that it is nicer to be a king in MMORPG land than a peasant in this one. But someday will it be nicer to be a king in MMORPG land than a king in real life?

Of course, people will always have to unplug from their stories to visit the bathroom and the refrigerator. But interactive fictions may become so appealing that we will be loath to leave them behind. This is something that the relentlessly optimistic *Star Trek* series never quite got right. The holodeck is, like the hydrogen bomb, a technology with hideous destructive potential. If you had a walk-in closet where you always got to do the thing you most wanted to do—from saving the world to mastering your harem—why would you ever come out? Why would you ever want to stop being god?

Humans evolved to crave story. This craving has, on the whole, been a good thing for us. Stories give us pleasure and instruction. They simulate worlds so we can live better in this one. They help bind us into communities and define us as cultures. Stories have been a great boon to our species.

But are they becoming a weakness? There's an analogy to be made between our craving for story and our craving for food. A tendency to overeat served our ancestors well when food shortages were a predictable part of life. But now that we modern desk jockeys are awash in cheap grease and corn syrup, overeating is more likely to fatten us up and kill us young. Likewise, it could be that an intense greed for story was healthy for our ancestors but has some harmful consequences in a world where books, MP3 players, TVs, and iPhones make story omnipresent—and where we have, in romance novels and television shows such as *Jersey Shore,* something like the story equivalent of deep-fried Twinkies. I think the literary scholar Brian Boyd is right to wonder if overconsuming in a world awash with junk story could lead to something like a "mental diabetes epidemic."

Similarly, as digital technology evolves, our stories—ubiquitous, immersive, interactive—may become dangerously attractive. The real threat isn't that story will fade out of human life in the future; it's that story will take it over completely.

Maybe we can avoid this fate. Maybe, like disciplined dieters, we can make nutritious choices and avoid gorging on story. In that spirit, here are some modest suggestions based on the research in this book.

Read fiction and watch it. It will make you more empathic and better able to navigate life's dilemmas.

Don't let moralists tell you that fiction degrades society's moral fabric. On the contrary, even the pulpiest fare usually pulls us together around common values.

Remember that we are, by nature, suckers for story. When emotionally absorbed in character and plot, we are easy to mold and manipulate.

Revel in the power of stories to change the world (think *Uncle Tom's Cabin*), but guard against it, too (think *The Birth of a Nation*).

Soccer practice and violin lessons are nice, but don't schedule away your child's time in Neverland—it is a vital part of healthy development.

Allow yourself to daydream. Daydreams are our own little stories: they help us learn from the past and plan for the future.

Recognize when your inner storyteller is locked in overdrive: be skeptical of conspiracy theories, your own blog posts, and self-exculpatory accounts of spats with spouses and coworkers.

If you are a doubter, try to be more tolerant of the myths—national and religious—that help tie culture together. Or at the very least, try to be less celebratory of their demise.

The next time a critic says that the novel is dying from lack of novelty, just yawn. People don't go to story land because they want something startlingly new; they go because they want the old comforts of the universal story grammar.

Don't despair for story's future or turn curmudgeonly over the rise of video games or reality TV. The way we experience story will evolve, but as storytelling animals, we will no more give it up than start walking on all fours.

Rejoice in the fantastic improbability of the twisting evolutionary path that made us creatures of story—that gave us all the gaudy, joyful dynamism of the stories we tell. And realize, most importantly, that understanding the power of storytelling—where it comes from and why it matters—can never diminish your experience of it. Go get lost in a novel. You'll see.

Acknowledgments

Thanks to all the individuals and organizations that gave me permission to use the images in this book. Thanks also to Frederic Spotts for patiently explaining copyright issues pertaining to Hitler's paintings. Finally, special thanks to Wikimedia Commons for providing searchable access to a huge trove of digital images.

I'm grateful to the scholars and scientists who offered advice on the book as a whole or on individual chapters: Brian Boyd, Joseph Carroll, Edward Castronova, Sam Fee, Michael Gazzaniga, and Katja Valli. My mother and father, Marcia and Jon, as well as my brother, Robert, also gave sound advice and encouragement. Tiffani, my wife, deserves thanks not only for commenting on drafts, but also for putting up with me when I was obsessed with this book.

I thank the librarians at Washington and Jefferson College library, especially Rachel Bolden and Alexis Rittenberger. They ordered me many scores of interlibrary loan books and never sicced the library police on me, even when I deserved it.

I am deeply indebted to my talented, hard-working, polymathically informed editor, Amanda Cook. Amanda collaborated with me in everything from the book's overarching design to the texture of its sentences. No book is perfect, but without Amanda's ear for language and instinct for storytelling, this one would have been much less so. Others on the Houghton Mifflin Harcourt team also deserve thanks for their skill and diligence, including editorial assistant Ashley Gilliam, who worked closely with me on the images in this volume; book designer Brian Moore; and copy editor extraordinaire Barbara Jatkola.

I feel very fortunate to have, in Max Brockman, such a capable and gifted agent. Max was there at the beginning of this project, helping me shape an amorphous clump of ideas into the beginnings of a book. He has also been a reliable source of comradeship and wise council.

Since they were small, my daughters have invited me into their pretend worlds as a sort of participant-observer — playing the roles of princes, Ken dolls, and ogres. Playing with my girls has taught me as much about story as I've ever learned in books. Thanks, Abigail. Thanks, Annabel.

Notes

PREFACE

xi *researchers from Plymouth:* BBC News 2003.

xii *Ssssss:* Elmo et al. 2002.

xii *"'the infinite monkey theory'":* BBC News 2003.

   "a less than infinite": Tanaka 2010.

xiv Homo fictus: E. M. Forster uses this term in *Aspects of the Novel* (1955, p. 55) to describe literary characters. See also Niles 1999.

1. THE WITCHERY OF STORY

1 *The Witchery of Story:* This phrase comes from Sara Cone Bryant's book *How to Tell Stories to Children* (Boston: Houghton Mifflin, 1905), p. 8.

3 *"Under Coffin's watchful eye":* Philbrick 2000, p. xii.

4 *"willing suspension":* Samuel Taylor Coleridge, *Biographia Literaria* (1817; repr., New York: Leavitt, Lord, 1834), p. 174.

8 *50 percent of Americans:* National Endowment for the Arts 2008.

   twenty minutes per day: Bureau of Labor Statistics 2009.

   more time in TV land: Shaffer et al. 2006, p. 623.

9 *five hours per day:* Motion Picture Association of America 2006. This estimate does not account for time spent downloading fictions from websites such as Hulu and YouTube.

   five hours of music: Levitin 2008, p. 3.

10 *storylike dreams:* Solms 2003.

   almost all night: Dream researcher Owen Flanagan thinks that we probably dream all night long (Flanagan 2000, p. 10).

11   *daydreaming is the mind's:* Klinger 2009.

Clever scientific studies: Ibid.

we spend about half: Using different methods, both Klinger 2009 and Killingsworth and Gilbert 2010 reached this conclusion.

13   *"We make movies":* Blaustein 2000.

14   New York Times Magazine *article:* Baker 2011.

15   *"fictional screen media":* See Bryant and Oliver 2009.

Business executives: Guber 2011.

16   *the role of story in court:* Malcolm 2010.

much good journalism: Wolfe 1975.

18   *gossipy stories:* Dunbar 1996; Norrick 2007.

2. THE RIDDLE OF FICTION

23   *romps, riots, and revels:* For overviews of the developmental psychology of children's pretend play, see Bloom 2004; Bjorklund and Pellegrini 2002; Boyd 2009; Gopnik 2009; Harris 2000; Singer and Singer 1990; Sutton-Smith 1997; Taylor 1999; Weisberg 2009.

Auschwitz: Eisen 1988; Walton 1990, p. 12.

24   *riddle of fiction:* For more, see Boyd 2009; Dutton 2009; Dissanayake 1995, 2000; Pinker 1997, 2002, 2007; Bloom 2010; Zunshine 2006; Knapp 2008; Wilson 1998, chapter 10. For a collection of major contributions, see Boyd, Carroll, and Gottschall 2010.

marvel of bioengineering: For facts and philosophical meditation on the human hand, see Napier 1993; Bell 1852; Tallis 2003; Wilson 1998.

25   *Clay bison:* See Breuil 1979; Bégouën et al. 2009.

27   *evolutionary source of story:* Darwin 1897; Miller 2001; Dutton 2009.

"a work of art": Boyd 2009, p. 15.

28   *vicarious experience:* Sugiyama 2005.

social glue: Dissanayake 1995, 2000.

"Real art creates": Gardner 1978, p. 125.

These and other theories: For the theory that art helps us maintain cognitive order or mental homeostasis, see Wilson 1998, pp. 210–37; Carroll 2008, pp. 119–28; Damasio 2010, pp. 294–97.

"Cocaine": Kessel 2010, p. 657.

29   *"for kicks":* Ibid., p. 672.

They are side effects: This view is most commonly associated with Pinker 1997, 2002, but Pinker actually thinks that fiction, unlike other art forms, may have an evolutionary purpose. Paul Bloom (2010) has strongly argued that fiction is an evolutionary side effect.

30   *deeply unsatisfactory:* See, for example, Boyd 2009.

32   *"Whatever else":* Paley 1988, p. 6.
33   *only one thing:* Bruner 2002, p. 23.
      "Will you tell me a story?": Sutton-Smith 1986.
34   *360 stories:* Appleyard 1990. For children's storytelling generally, see Engel 1995.
      "The typical actions": Sutton-Smith, 1997, pp. 160–61.
35   *"Where's the baby":* Paley 2004, p. 60.
      "Pretend you're a frog": Ibid., p. 30.
38   *"Let the boys be robbers":* Paley 1984, p. 116.
39   *reliable sex differences:* Konner 2010.
      Dozens of studies: For a review of studies, see Konner 2010. See also Geary 1998;
      Bjorklund and Pellegrini 2002.
      seventeenth month of life: Konner 2010, chap. 19.
      "Most of the time": Singer and Singer 1990, p. 78.
      girl play only seems: Paley 1984, p. 58.
40   *"affected girls show":* Konner 2010, p. 270.
      a bad guy named Lurky: Paley 1984, p. 84.
41   *"a muddle out of which":* The quote is a description of Piaget's views by the psy-
      chologist J. A. Appleyard (1990, p. 11).
      the work of children: For reviews of children's pretend play, see Bloom 2004;
      Bjorklund and Pellegrini 2002; Gopnik 2009; Singer and Singer 1990; Boyd
      2009; Sutton-Smith 1997; Taylor 1999; Weisberg 2009.
      have never found: See Konner 2010, p. 264; Wood and Eagly 2002.
      "I blush": Poe 1975, p. 224.
42   *"A man once slaughtered":* Tatar 2003, p. 247.
44   *a critic counted:* Russell 1991, p. 74.
      in a different study: Davies et al. 2004.

3. HELL IS STORY-FRIENDLY

51   Mr Bailey, Grocer: Thanks to Joseph Carroll for bringing this example to my at-
      tention.
52   *"Conflict is the fundamental":* Burroway 2003, p. 82.
      "Hell is story-friendly": Baxter 1997, p. 133.
53   *"There is very little":* James 2007, p. 257.
54   *"Margaritomancy!":* Joyce 1999, p. 281.
55   *"nothing much happens":* Quoted in Baxter 1997, p. 226.
      master themes: For a recent treatment of the topic, with reviews of previous
      scholarship, see Booker 2004.
56   *Navy fighter pilots:* For details of flight training, see Waller 1999.
57   *story is where people:* For variations on this idea, see Boyd 2009; Pinker 1997,
      2002; Sugiyama 2005.

"Literature offers feelings": Burroway 2003, p. 74. Italics in original.

58 *flight simulators of human social life:* Oatley, "The Mind's Flight Simulator," 2008.

59 *Italian neuroscientists:* For basic overviews of the research, see Iacoboni 2008; Rizzolatti, Sinigaglia, and Anderson 2008; Ramachandran 2011. For skepticism, see Hickok 2009; Dinstein et al. 2008.

60 *mirror neurons may help:* For the original study of imitation in newborns, see Meltzoff and Moore 1977. For infant imitation and mirror neurons, see Meltzoff and Decety 2003.

61 *"because mirror neurons":* Iacoboni 2008, p. 4.

Other scientists are more wary: For instance, Ilan Dinstein and his colleagues allow that "mirror neurons are exceptionally interesting neurons, which may underlie certain social capabilities in both animals and humans," but point out that the "'human mirror system' in particular has been characterized by much speculation and relatively little hard evidence" (Dinstein et al. 2008, p. 17).

stories affect us physically: Nell 1988.

62 *"media equals real life":* Reeves and Nass 2003.

in a Dartmouth brain lab: Krendl et al. 2006.

63 *they tend to respond:* Slater et al. 2006.

"What this means": Jabbi, Bastiaansen, and Keysers 2008. For a similar fMRI study of fictional response, see Speer et al. 2009.

64 *"cells that fire together":* The Canadian psychologist Donald Hebb, quoted in Ledoux 2003, p. 79. Daniel Goleman writes, "Simulating an act is, in the brain, the same as performing it, except that the actual execution is somehow blocked" (Goleman 2006, pp. 41–42).

stories equip us: Pinker 1997, chap. 8.

fiction can make: See Fodor 1998.

Almost none of the details: On the fragmentary nature of fictional recollection, see Bayard 2007.

65 implicit *memory:* See Schachter 1996, 2001.

"realistic rehearsal": Valli and Revonsuo 2009, p. 11.

66 *researchers are simply not:* On the application of scientific methods to literary questions, see Gottschall 2008.

flight simulators work: Lehrer 2010, pp. 252–53.

fiction readers had better: Mar et al. 2006. For overviews of this research, see Oatley, "The Mind's Flight Simulator," 2008; Oatley, "The Science of Fiction," 2008; Oatley 2011.

In a second test: Mar, Oatley, and Peterson 2009. For Oatley's quote and a more thorough description of Mar, Oatley, and Peterson's methods and findings, see Oatley 2011. See also Keen 2007. Pinker 2011 argues that "it seems likely that fiction . . . helps to expand people's circle of sympathy."

4. NIGHT STORY

70 *"vivid and continuous dream[s]"*: Gardner 1983, p. 32.

    "sensorimotor hallucinations": Koch 2010, p. 16

71 *"Freud was determined"*: Crews 2006, p. 24.

72 *"fortune cookie" model:* Hobson 2002, p. 64.

    "The manifest dream": Ibid., p. 151.

73 *"We dream to forget"*: Crick and Mitchison 1983, p. 111.

    "Our dreams were not": Flanagan 2000, p. 24.

74 *dreams are mainly:* Valli and Revonsuo 2009, p. 25. See also Franklin and Zyphur 2005, p. 64; Talbot 2009, p. 47.

    "No random process": Revonsuo 2003, p. 278.

77 *"dreams of actions"*: Jouvet 1999, p. 92.

78 *rats probably dream:* Ji and Wilson 2007.

79 *"[In dreams] waves"*: Quoted in Revonsuo 2000, p. 898.

81 *In a 2009 review:* Valli and Revonsuo 2009.

    REM behavior disorder: For overviews of RBD, see Valli and Revonsuo 2009; Revonsuo 2000.

82 *"the stage on which"*: Paley 1988, p. 65.

    "are so rare": Revonsuo 2003, p. 280.

    Other universal themes: Revonsuo 2003, p. 288.

83 *studied dream reports:* Valli and Revonsuo 2009.

84 *"We experience a dream"*: Quoted in Rock 2004, p. 1.

    "When you consider": Franklin and Zyphur 2005, p. 73. For another adaptationist perspective on dreams, see McNamara 2004.

85 *"If you attempt"*: Fajans, "How You Steer a Bicycle."

86 *"It is difficult"*: Hunt 2003, p. 166.

    dreams are reasonably realistic: Franklin and Zyphur 2005, p. 64.

    "relevant, reasonable": Valli and Revonsuo 2009, p. 25.

    "We are for the first time": Revonsuo 2003, p. 294.

5. THE MIND IS A STORYTELLER

91 *"the central mystery"*: Nettle 2001, p. 117.

    turn flat characters round: Forster 1955, pp. 67–78.

92 *"We of the craft"*: Quoted in Gardiner 1836, p. 87.

93 *"It is hard to avoid""*: Nettle 2001, p. 147.

    "The Western literary tradition": Allen 2004, p. ix.

    massive study: Ludwig 1996.

    underlying genetic component: See Nettle 2001; Jamison 1993.

94 *In his memoir:* King 2000, pp. 90–91.

95   *dangerous experimental procedure:* For details on split-brain operations and research, see Gazzaniga 2000; Gazzaniga, *Human,* 2008; Gazzaniga, "Forty-Five Years of Split-Brain Research," 2008; Funnel, Corbalis, and Gazzaniga 2000.

96   *left brain is specialized:* Gazzaniga, *Human,* 2008, p. 13.

"the interpreter": For overviews, see Gazzaniga 2000; Gazzaniga, *Human,* 2008; Gazzaniga, "Forty-Five Years of Split-Brain Research," 2008.

97   *In one experiment:* Gazzaniga, *Human,* 2008, pp. 294–95.

101 *"You have been":* Doyle 1904, p. 8.

With great relish: Ibid., p. 22.

103 *The human mind is tuned:* Hood 2009.

104 *"Human beings like stories":* Wallis 2007, p. 69.

105 *a very limited ability:* See Haven 2007. For harmful consequences of our tendency to impose story patterns on random information, see Taleb 2008.

This point is: Heider and Simmel 1944.

108 *"rock-jawed certainty":* Hirstein 2006, p. 8. See also Hirstein 2009.

"confabulatory genius" who "must literally": Sacks 1985, p. 110. Italics in original.

When asked to move: Hirstein 2006, pp. 135–36. For an overview of confabulatory syndromes, see Hirstein 2006.

109 *One of the first:* Maier 1931.

110 *In a more recent study:* Described in Wheatley 2009.

lies, honestly told: For another fascinating example of confabulation in ordinary people, see Johansson et al. 2005.

111 *In his documentary:* Jones 2007.

112 *Documentarians . . . recently followed: New World Order,* directed by Luke Meyer and Andrew Neel (New York: Disinformation, 2009).

113 *one million listeners:* "Angry in America: Inside Alex Jones' World," ABC News/ Nightline, September 2, 2010, http://abcnews.go.com/Nightline/alex-jones-day-life-libertarian-radio-host/story?id=10891854&page=2 (accessed September 18, 2010).

vampiristic extraterrestrial lizard people: Icke 1999.

114 *"a walking compendium":* Olmsted 2009, p. 197.

115 *gunned down by Stephen King:* See Lightfoot 2001.

Scripps Howard poll: Hargrove 2006. For debunking of 9/11 conspiracies, see Dunbar and Regan 2006.

Obama is a stealth Muslim: "Growing Number of Americans Say Obama Is a Muslim," Pew Research Center, August 19, 2010, http://pewresearch.org/ pubs/1701/poll-obama-muslim-christian-church-out-of-politics-political-leaders-religious (accessed September 18, 2010); "Time Magazine/ABT SRBI—August 16–17, 2010 Survey: Final Data," http://www.srbi.com/TimePoll5122-Final%20Report-2010-08-18.pdf (accessed September 18, 2010).

Obama was not born: "The President, Congress and Deficit Battles: April 15–20,

2011," CBS News/NYT Polls, http://www.cbsnews.com/stories/2011/04/21/ politics/main20056282.shtml?tag=contentMain;contentBody (accessed July 21, 2011). Another 25 percent of Republicans had no opinion on the question.

Obama is the Antichrist: "Quarter of Republicans Think Obama May Be the Anti-Christ," LiveScience, March 25, 2010, http://www.livescience.com/culture/ obama-anti-christ-100325.html (accessed September 18, 2010).

"Conspiracy theories originate": Aaronovitch 2010, p. 338.

6. THE MORAL OF THE STORY

117 *one writer puts:* Jacobs 2008, p. 8.

118 *if you want a message:* For an overview of the research, see Haven 2007.

119 *"ungoded":* Quoted in Gardner 1978, p. 9.

world . . . is getting: Johnson 2008. On Americans getting more religious, see Kagan 2009, p. 86.

120 *an ancient story:* Genesis 17:10–14.

121 *most current evolutionary thinkers:* See Bulbulia et al. 2008; Voland and Schiefenhovel 2009; Boyer 2002; Dawkins 2006; Dennett 2007.

Daniel Dennett and Richard Dawkins: See Dennett 2007; Dawkins 2006.

"a virus of the mind": Dawkins 2004, p. 128.

religion emerged: Wilson 2003. Wilson's idea is based on the concept of group selection. The idea that natural selection can work at the group level has been, for almost fifty years, among the most controversial ideas in evolutionary biology. For an overview of the controversy and a description of a recent renaissance in group selection theory, see Wilson and Wilson 2007; Wilson 2007.

122 *"Religion is a unified system":* Durkheim 2008, p. 46.

"is to bind people": Wade 2009, p. 58.

"elements of religion": Wilson 2008, p. 27. Italics in original.

"a weapon in": Jager 1869, p. 119.

124 *"They would make":* Quoted in Zinn 2003, p. 1

Revisionist historians: See Zinn 2003; Loewen 1995.

spinners of countermyths: For a critique of historical revisionism and a defense of the basic correctness of traditional American history, see Schweikart and Allen 2007.

128 *exposes people:* Haidt 2006, pp. 20–21.

"a dead-but-living": Stone 2008, p. 181.

129 *the philosopher David Hume:* Hume 2010, p. 283.

"imaginative resistance": Gendler 2000.

130 *"two men kill":* Pinker 2011.

Plato banished poets: Plato 2003, p. 85.

132 *200,000 violent acts:* Linn 2008, p. 158.

"a highly traditionalist": Ong 1982, p. 41. On the conservatism of traditional art generally, see Dissanayake 1995, 2000.

"great fiction is subversive": Bruner 2002, p. 11.

fiction is . . . deeply moral: Tolstoy 1899; Gardner 1978.

133   *"death of the antagonist"*: Baxter 1997, p. 176.

134   *"The truth is"*: Johnson 2005, pp. 188–89.

William Flesch thinks: Flesch 2007.

In a series of papers: Carroll et al. 2009, 2012. For similar arguments, see Boyd 2009, pp. 196–97.

135   *"moral overtones"*: Elkind 2007, p. 162.

collision of evil and good: See Paley 1988.

reviewed dozens: Hakemulder 2000.

136   *"this is patently"*: Appel 2008, pp. 65–66.

It choreographs how: Nell 1988.

138   *global village:* McLuhan 1962.

"is essentially serious": Gardner 1978, p. 6.

7. INK PEOPLE CHANGE THE WORLD

139   *August Kubizek relates:* Kubizek 2006, pp. 116–17.

140   *As a young man:* Details of Hitler's early life and his *Rienzi* epiphany are discussed in Kohler 2000; Kershaw 1998; Fest 1974; Spotts 2003; Kubizek 2006.

141   *"was rejected"*: Spotts 2003, p. 138. In 1936, a selection of Hitler's paintings was published as a coffee-table book titled *Adolf Hitler: Bilder aus dem Leben des Führers* (Hamburg: Cigaretten Bilderdienst, 1936).

variously spelled: Shirer 1990, pp. 6–7.

"Hitler is one": Kershaw 1998, p. xx.

the Rienzi episode seems: See Kohler 2000, pp. 25–26; Spotts 2003, pp. 226–27.

143   *"grand solution"*: Nicholson 2007, pp. 165–66.

Hitler "'lived' Wagner's work": Quoted in Kohler 2000, p. 293.

144   *"For the Master"*: Fest 1974, p. 56.

"whoever wants to understand": Quoted in Viereck 1981, p. 132.

145   *"as much effect"*: Encyclopaedia Britannica, 11th ed., s.v. "poetry."

146   *"blacksmith's hammer"*: Stowe 2007, p. 357.

147   *"So you're the little"*: Quoted in Stowe and Stowe 1911, p. 223.

"exerted a momentous impact": Weinstein 2004, p. 2.

"In Britain": Johnson 1997, p. 417.

resurrected the defunct: Clooney 2002, pp. 277–94.

Jaws . . . depressed: Gerrig 1993, pp. 16–17.

148   *"the grisly inheritance"*: Hitchens 2010, p. 98.

"Would Alexander": Richardson 1812, p. 315.

149 *fiction can mislead:* Appel and Richter 2007. See also Gerrig and Prentice 1991; Marsh, Meade, and Roediger 2003.

"*the stronger the infection*": Tolstoy 1899, p. 133.

have been traumatized: Cantor 2009.

150 "*Researchers have repeatedly*": Mar and Oatley 2008, p. 182. See also Green and Brock 2000.

more effective at changing: Green and Donahue 2009; Green, Garst, and Brock 2004.

a single viewing: Eyal and Kunkel 2008.

The effects of violence: For reviews of research on the effects of violent media, see Anderson et al. 2010. For skeptical views, see Jones 2002; Schechter 2005. For the relationship between prosocial behavior and prosocial games, see Greitemeyer and Osswald 2010.

151 *how we view out-groups:* For research on fiction influencing racial attitudes, see Mastro 2009; Rosko-Ewoldsen et al. 2009. For research on fiction influencing gender attitudes, see Smith and Granados 2009.

fiction writers mix: Maugham 1969, p. 7.

A related explanation: Green and Brock 2000.

152 *Stories change our beliefs:* Djikic et al., 2009. See also Mar, Djikic, and Oatley 2008, p. 133.

153 "*Hitler's interest in the arts*": Spotts 2003, p. xii

an ecstasy of book burning: United States Holocaust Museum, "Book Burning."

154 "*the greatest actor*": Quoted in Spotts 2003, p. 43.

"*Hitler was one*": Quoted in Spotts 2003, p. 56.

155 "*Where they burn books*": Quoted in Metaxas 2010, p. 162.

8. LIFE STORIES

156 "*How old was I*": Conversation between the old man, Santiago, and his young apprentice, Manolin.

The day before he got fired: Carr 2008, pp. 3–8, 10, 12.

158 *Frey describes:* Frey 2008.

a masterpiece of debunking: Smoking Gun 2006.

159 "*will probably be remembered*": Yagoda 2009, p. 246.

"*My literary lineage*": Quoted in ibid., p. 22.

160 *Another celebrated:* Carter 1991. See also Barra 2001.

"*virulent segregationist*": Barra 2001.

161 *a "personal myth"*: McAdams 1993. See also McAdams 2001, 2008.

162 "*Memory . . . is never true*": Hemingway 1960, p. 84.

Marie G. reported: Bernheim 1889, pp. 164–66; Lynn, Matthews, and Barnes 2009.

"*before God and man*": Bernheim 1889, p. 165.

"*flashbulb memories*": Brown and Kulik 1977.

163 *researchers asked people:* Neisser and Harsch 1992.

"*For a quarter*": French, Garry, and Loftus 2009, pp. 37–38.

164 *The research shows:* French, Garry, and Loftus 2009; Greenberg 2004.

"*I was in Florida*": CNN 2001.

The headline: Greenberg 2004, p. 364.

165 *73 percent of research subjects:* Greenberg 2005, p. 78.

many British people: Ost et al. 2008.

"*an intelligent woman*": Bernheim 1889, p. 164.

166 "*incontestable reality*": Ibid.

Memory continued to be seen: For early work on false memory, see Bartlett 1932.

167 "*the great sex panic*": Crews 2006, p. 154. This book provides a skeptical analysis and history of recovered memory.

In a classic experiment: See Loftus and Pickrell 1995; French, Garry, and Loftus 2009.

This study was among: For overviews of the false-memory research described here, see Brainerd and Reyna 2005; Schachter 1996, 2001; Bernstein, Godfrey, and Loftus 2009.

168 *every bit as confident:* Brainerd and Reyna 2005, pp. 20, 409.

For example, in one study: Schachter 2001, p. 3. For another study of ordinary misremembering, see Conway et al. 1996.

169 *Pieces of that memory:* See Young and Saver 2001; Schachter 2001, pp. 85–87.

170 *it just doesn't work:* Marcus 2008, chapter 2.

"*serve many masters*": Bruner 2002, p. 23.

"*unreliable, self-serving historian*": Tavris and Aronson 2007, p. 6.

Even truly awful people: The writer and corrections officer Rory Miller writes that most criminals convince themselves that they are the real good guys and the real victims (Miller 2008, p. 100).

"*Annie Wilkes*": King 2000, p. 190.

fold it into a narrative: Baumeister 1997; Kurzban 2010. For a review of the self-exculpatory bias, see Pinker 2011, chapter 8.

the "Great Hypocrisy": Ibid.

171 "*I see myself*": Quoted in Baumeister 1997, p. 49.

172 *photographic distortion:* Tsai 2007.

"*70% thought*": Gilovich 1991, p. 77. Italics in original.

173 *For example:* Taylor 1989, p. 10; Gilovich 1991, p. 77.

College students generally believe: Taylor 1989. See also Mele 2001.

Lake Woebegone effect: Pronin, Linn, and Ross 2002.

improving with age: Wilson and Ross 2000.

174 *a "farce" and an "agreeable fiction":* Fine 2006, pp. 6, 25.

   Self-aggrandizement starts: Taylor 1989, p. 200.

   a healthy mind: Ibid., p. xi.

175 *"The truth is depressing":* Hirstein 2006, p. 237.

   depression frequently stems: Crossley 2000, p. 57.

   recent review article: Shedler 2010.

   a kind of script doctor: See Spence 1984.

9. THE FUTURE OF STORY

177 *who killed poetry:* See Epstein 1988; Fenza 2006.

   English departments have been: Gottschall 2008.

178 *made more money:* Baker 2010.

   "I come to": Shields 2010, p. 175.

   love to wallow: For a recent example of wallowing in the novel's obsolescence, see Shields 2010. James Wood (2008), while not so pessimistic, wallows along similar lines.

   numbers trending up: For statistics compiled by R. R. Bowker, the company that creates the Books in Print database and assigns ISBNs to new books, see "New Book Titles and Editions, 2002–2009," http://www.bowkerinfo.com/bowker/IndustryStats2010.pdf (accessed November 21, 2010).

   a new novel is published: Miller 2004. This figure excludes print-on-demand and vanity press novels.

181 *The 2010 release:* Bradley and DuBois 2010.

   "most widely disseminated": Bradley 2009, p. xiii.

182 *we are living through:* Bissell 2010.

   the writer/director: Bland 2010.

183 *Together with teams:* William Booth, "Reality Is Only an Illusion, Writers Say," *Washington Post,* August 10, 2004, http://www.washingtonpost.com/wp-dyn/articles/A53032-2004Aug9.html (accessed March 4, 2011).

184 *This enraged Junk's friend: The Ultimate Fighter* 2009.

186 *"The future looks bleak":* Quoted in Castronova 2007, p. 46.

187 *Ethan and his friends:* Gilsdorf 2009, pp. 88, 101, 104, 105.

192 *anthropologists write:* For one ambitious example and references to other studies, see Bainbridge 2010.

193 *As one player put it:* Kelly 2004, pp. 11, 71.

   "tapestry of myths": Bainbridge 2010, p. 14.

194 *greatest mass migration:* Castronova 2007.

195 *twenty to thirty hours:* Kelly 2004, p. 13.

   most satisfying friendships: Meadows 2008, p. 50.

20 percent consider: Castronova 2007, p. 13.

196   *daily vacation:* Quoted in Castronova 2006, p. 75.

"In the lands": "The Chaos Warhost: The Chosen," War Vault Wiki, Warhammer Online, http://warhammervault.ign.com/wiki/index.php?title=Chosen&oldid =1997 (accessed July 23, 2011).

reality is broken: McGonigal 2011.

198   *"mental diabetes epidemic":* Brian Boyd, personal communication, October 3, 2010.

Bibliography

Aaronovitch, David. *Voodoo Histories: The Role of the Conspiracy Theory in Shaping Modern History.* New York: Riverhead, 2010.

Allen, Brooke. *Artistic License: Three Centuries of Good Writing and Bad Behavior.* New York: Ivan R. Dee, 2004.

Anderson, Craig A., Akiko Shibuya, Nobuko Ihori, Edward L. Swing, Brad J. Bushman, Akira Sakamoto, Hannah R. Rothstein, and Muniba Saleem. "Violent Video Game Effects on Aggression, Empathy, and Prosocial Behavior in Eastern and Western Countries: A Meta-Analytic Review." *Psychological Bulletin* 136 (2010): 151–73.

Appel, Markus. "Fictional Narratives Cultivate Just-World Beliefs." *Journal of Communication* 58 (2008): 62–83.

Appel, Markus, and Tobias Richter. "Persuasive Effects of Fictional Narratives Increase over Time." *Media Psychology* 10 (2007): 113–34.

Appleyard, J. A. *Becoming a Reader: The Experience of Fiction from Childhood to Adulthood.* Cambridge: Cambridge University Press, 1990.

Bainbridge, William. *The Warcraft Civilization: Social Science in a Virtual World.* Cambridge, MA: MIT Press, 2010.

Baker, Katie. "The XX Blitz: Why Sunday Night Football Is the Third-Most-Popular Program on Television—Among Women." *New York Times Magazine,* January 30, 2011.

Baker, Nicholson. "Painkiller Deathstreak." *The New Yorker,* August 9, 2010.

Barra, Allen. "The Education of Little Fraud." Salon, December 20, 2001. http://dir.salon.com/story/books/feature/2001/12/20/carter/index.html. Accessed February 16, 2011.

Bartlett, Frederic. *Remembering: A Study in Experimental and Social Psychology.* Cambridge: Cambridge University Press, 1932.

Baumeister, Roy. *Evil: Inside Human Violence and Cruelty.* New York: Henry Holt, 1997.

Baxter, Charles. *Burning Down the House: Essays on Fiction.* St. Paul: Graywolf, 1997.

Bayard, Pierre. *How to Talk About Books You Haven't Read.* London: Bloomsbury, 2007.

BBC News. "No Words to Describe Monkeys' Play," May 9, 2003. http://news.bbc .co.uk/2/hi/3013959.stm. Accessed August 30, 2010.

Bégouën, Robert, Carole Fritz, Gilles Tosello, Jean Clottes, Andreas Pastoors, and François Faist. *Le Sanctuaire secret des bisons.* Paris: Somogy, 2009.

Bell, Charles. *The Hand: Its Mechanisms and Vital Endowments, as Evincing Design.* London: John Murray, 1852.

Bernheim, Hippolyte. *Suggestive Therapeutics: A Treatise on the Nature and Uses of Hypnotism.* New York: G. P. Putnam's Sons, 1889.

Bernstein, Daniel, Ryan Godfrey, and Elizabeth Loftus. "False Memories: The Role of Plausibility and Autobiographical Belief." In *Handbook of Imagination and Mental Simulation,* edited by Keith Markman, William Klein, and Julie Suhr, 89–102. New York: Psychology Press, 2009.

Bissell, Tom. *Extra Lives: Why Video Games Matter.* New York: Pantheon, 2010.

Bjorklund, David, and Anthony Pellegrini. *The Origins of Human Nature: Evolutionary Developmental Psychology.* Washington, DC: American Psychological Association, 2002.

Bland, Archie. "Control Freak: Will David Cage's 'Heavy Rain' Video Game Push Our Buttons?" *Independent,* February 21, 2010.

Blaustein, Barry. *Beyond the Mat.* Los Angeles: Lion's Gate Films, 2000. Film.

Bloom, Paul. *Descartes' Baby: How the Science of Child Development Explains What Makes Us Human.* New York: Basic, 2004.

———. *How Pleasure Works: The New Science of Why We Like What We Like.* New York: Norton, 2010.

Blume, Michael. "The Reproductive Benefits of Religious Affiliation." In *The Biological Evolution of Religious Mind and Behavior,* edited by Eckart Voland and Wulf Schiefenhovel. Dordrecht, Germany: Springer, 2009.

Booker, Christopher. *The Seven Basic Plots.* New York: Continuum, 2004.

Boyd, Brian. *On the Origin of Stories: Evolution, Cognition, Fiction.* Cambridge, MA: Harvard University Press, 2009.

Boyd, Brian, Joseph Carroll, and Jonathan Gottschall. *Evolution, Literature, and Film: A Reader.* New York: Columbia University Press, 2010.

Boyer, Pascal. *Religion Explained.* New York: Basic, 2002.

Bradley, Adam. *Book of Rhymes: The Poetics of Hip Hop.* New York: Basic, 2009.

Bradley, Adam, and Andrew DuBois, eds. *The Anthology of Rap.* New Haven, CT: Yale University Press, 2010.

Brainerd, Charles, and Valerie Reyna. *The Science of False Memory.* Oxford: Oxford University Press, 2005.

Breuil, Abbé Henri. *Four Hundred Centuries of Cave Art.* New York: Hacker Art Books, 1979.

Brown, Roger, and James Kulik. "Flashbulb Memories." *Cognition* 5 (1977): 73–99.

Bruner, Jerome. *Making Stories: Law, Literature, Life.* New York: Farrar, Straus and Giroux, 2002.

Bryant, Jennings, and Mary Beth Oliver, eds. *Media Effects: Advances in Theory and Research.* 3rd ed. New York: Routledge, 2009.

Bulbulia, Joseph, Richard Sosis, Erica Harris, and Russell Genet, eds. *The Evolution of Religion: Studies, Theories, and Critiques.* Santa Margarita, CA: Collins Foundation, 2008.

Bureau of Labor Statistics. "American Time Use Survey—2009 Results." http://www.bls.gov/news.release/archives/atus_06222010.pdf. Accessed August 30, 2010.

Burroway, Janet. *Writing Fiction: A Guide to Narrative Craft.* 3rd ed. New York: Longman, 2003.

Cantor, Joanne. "Fright Reactions to Mass Media." In *Media Effects: Advances in Theory and Research,* 3rd ed., edited by Jennings Bryant and Mary Beth Oliver. New York: Routledge, 2009.

Carr, David. *The Night of the Gun.* New York: Simon and Schuster, 2008.

Carroll, Joseph. "An Evolutionary Paradigm for Literary Study." *Style* 42 (2008): 103–35.

Carroll, Joseph, Jonathan Gottschall, John Johnson, and Dan Kruger. "Human Nature in Nineteenth-Century British Novels: Doing the Math." *Philosophy and Literature* 33 (2009): 50–72.

———. *Graphing Jane Austen: The Evolutionary Basis of Literary Meaning.* NY: Palgrave, 2012.

———. "Paleolithic Politics in British Novels of the Nineteenth Century." In *Evolution, Literature, and Film: A Reader,* edited by Brian Boyd, Joseph Carroll, and Jonathan Gottschall. New York: Columbia University Press, 2010.

Carter, Dan. "The Transformation of a Klansman." *New York Times,* October 4, 1991.

Castronova, Edward. *Exodus to the Virtual World: How Online Fun Is Changing Reality.* New York: Palgrave, 2007.

———. *Synthetic Worlds: The Business and Culture of Online Games.* Chicago: University of Chicago Press, 2006.

Clooney, Nick. *The Movies That Changed Us.* New York: Atria, 2002.

CNN. "President Bush Holds Town Meeting," December 4, 2001. http://transcripts.cnn.com/TRANSCRIPTS/0112/04/se.04.html. Accessed August 30, 2010.

Conway, Martin, Alan Collins, Susan Gathercole, and Steven Anderson. "Recollections of True and False Autobiographical Memories." *Journal of Experimental Psychology* 125 (1996): 69–95.

Crews, Frederick. *Follies of the Wise: Dissenting Essays.* Emeryville, CA: Shoemaker Hoard, 2006.

Crick, Francis, and Graeme Mitchison. "The Function of Dream Sleep." *Nature* 304 (1983): 111–14.

Crossley, Michele. *Introducing Narrative Psychology: Self, Trauma, and the Construction of Meaning.* Buckingham, UK: Open University Press, 2000.

Damasio, Antonio. *Self Comes to Mind: Constructing the Conscious Brain.* New York: Pantheon, 2010.

Darwin, Charles. *The Descent of Man and Selection in Relation to Sex.* New York: D. Appleton, 1897. First published 1871.

Davies, P., L. Lee, A. Fox, and E. Fox. "Could Nursery Rhymes Cause Violent Behavior? A Comparison with Television Viewing." *Archives of Diseases of Childhood* 89 (2004): 1103–5.

Dawkins, Richard. *A Devil's Chaplain.* Boston: Mariner, 2004.

——. *The God Delusion.* Boston: Houghton Mifflin, 2006.

Dennett, Daniel. *Breaking the Spell: Religion as a Natural Phenomenon.* New York: Penguin, 2007.

Dinstein, Ilan, Cibu Thomas, Marlene Behrmann, and David Heeger. "A Mirror Up to Nature." *Current Biology* 18 (2008): 13–18.

Dissanayake, Ellen. *Art and Intimacy: How the Arts Began.* Seattle: University of Washington Press, 2000.

——. *Homo Aestheticus: Where Art Comes From and Why.* Seattle: University of Washington Press, 1995.

Djikic, Maja, Keith Oatley, Sara Zoeterman, and Jordan Peterson. "On Being Moved by Art: How Reading Fiction Transforms the Self." *Creativity Research Journal* 21 (2009): 24–29.

Doyle, A. C. *A Study in Scarlet, and, The Sign of the Four.* New York: Harper and Brothers, 1904. First published 1887.

Dunbar, David, and Brad Regan, eds. *Debunking 9-11 Myths.* New York: Hearst, 2006.

Dunbar, Robin. *Grooming, Gossip, and the Evolution of Language.* Cambridge, MA: Harvard University Press, 1996.

Durkheim, Émile. *The Elementary Forms of Religious Life.* Oxford: Oxford University Press, 2008. First published 1912.

Dutton, Denis. *The Art Instinct: Beauty, Pleasure, and Human Evolution.* New York: Bloomsbury, 2009.

Eisen, Greg. *Children and Play in the Holocaust.* Amherst, MA: University of Massachusetts Press, 1988.

Elkind, David. *The Power of Play: Learning What Comes Naturally.* New York: Da Capo, 2007.

Elmo, Gum, Heather, Holly, Mistletoe, and Rowan. *Notes Towards the Complete Works of Shakespeare*. Vivaria.net, 2002. http://www.vivaria.net/experiments/notes/publication/NOTES_EN.pdf. Accessed August 30, 2010.

Engel, Susan. *The Stories Children Tell: Making Sense of the Narratives of Childhood*. New York: Freeman, 1995.

Epstein, Joseph. "Who Killed Poetry?" *Commentary* 86 (1988): 13–20.

Eyal, Keren, and Dale Kunkel. "The Effects of Sex in Television Drama Shows on Emerging Adults' Sexual Attitudes and Moral Judgments." *Journal of Broadcasting and Electronic Media* 52 (2008): 161–81.

Fajans, Joel. "How You Steer a Bicycle." http://socrates.berkeley.edu/~fajans/Teaching/Steering.htm. Accessed August 30, 2010.

Fenza, David W. "Who Keeps Killing Poetry?" *Writer's Chronicle* 39 (2006): 1–10.

Fest, Joachim. *Hitler*. Translated by Richard Winston and Clara Winston. New York: Harcourt Brace Jovanovich, 1974.

Fine, Cordelia. *A Mind of Its Own: How Your Brain Distorts and Deceives*. New York: Norton, 2006.

Flanagan, Owen. *Dreaming Souls: Sleep, Dreams, and the Evolution of the Conscious Mind*. Oxford: Oxford University Press, 2000.

Flesch, William. *Comeuppance: Costly Signaling, Altruistic Punishment, and Other Biological Components of Fiction*. Cambridge, MA: Harvard University Press, 2007.

Fodor, Jerry. "The Trouble with Psychological Darwinism." *London Review of Books*, January 15, 1998.

Forster, E. M. *Aspects of the Novel*. New York: Mariner, 1955. First published 1927.

Franklin, Michael, and Michael Zyphur. "The Role of Dreams in the Evolution of the Mind." *Evolutionary Psychology* 3 (2005): 59–78.

French, Lauren, Maryanne Garry, and Elizabeth Loftus. "False Memories: A Kind of Confabulation in Non-Clinical Subjects." In *Confabulation: Views from Neuroscience, Psychiatry, Psychology, and Philosophy*, edited by William Hirstein, 33–66. Oxford: Oxford University Press, 2009.

Freud, Sigmund. *The Interpretation of Dreams*. 3rd ed. N.p.: Plain Label, 1911. First published 1900.

Frey, James. *A Million Little Pieces*. New York: Anchor Books, 2004.

Funnel, Margaret, Paul Corbalis, and Michael Gazzaniga. "Hemispheric Interactions and Specializations: Insights from the Split Brain." In *Handbook of Neuropsychology*, edited by Francois Boller, Jordan Grafman, and Giacomo Rizzolatti, 103–20. Amsterdam: Elsevier, 2000.

Gardiner, Marguerite. *Conversations of Lord Byron with the Countess of Blessington*. Philadelphia: E. A. Carey and Hart, 1836.

Gardner, John. *The Art of Fiction: Notes on Craft for Young Writers*. New York: Vintage, 1983.

———. *On Moral Fiction.* New York: Basic, 1978.

Gass, William. *Fiction and the Figures of Life.* New York: Godine, 1958.

Gazzaniga, Michael. "Forty-Five Years of Split-Brain Research and Still Going Strong." *Nature Reviews Neuroscience* 6 (2008): 653–59.

———. *Human.* New York: HarperCollins, 2008.

———. *The Mind's Past.* Berkeley: University of California Press, 2000.

Geary, David. *Male and Female: The Evolution of Human Sex Differences.* Washington, DC: American Psychological Association, 1998.

Gendler, Tamar. *Thought Experiment: On the Powers and Limits of Imaginary Cases.* New York: Garland, 2000.

Gerrig, Richard. *Experiencing Narrative Worlds: On the Psychological Activities of Reading.* New Haven, CT: Yale University Press, 1993.

Gerrig, Richard, and Deborah Prentice. "The Representation of Fictional Information." *Psychological Science* 2 (1991): 336–40.

Gilovich, Thomas. *How We Know What Isn't So.* New York: Macmillan, 1991.

Gilsdorf, Ethan. *Fantasy Freaks and Gaming Geeks.* Guilford, CT: Lyons, 2009.

Goleman, Daniel. *Social Intelligence: The Revolutionary New Science of Human Relationships.* New York: Bantam, 2006.

Gopnik, Alison. *The Philosophical Baby: What Children's Minds Tell Us About Truth, Love, and the Meaning of Life.* New York: Farrar, Straus and Giroux, 2009.

Gottschall, Jonathan. *Literature, Science, and a New Humanities.* New York: Palgrave, 2008.

Green, Melanie, and Timothy Brock. "The Role of Transportation in the Persuasiveness of Public Narratives." *Journal of Personality and Social Psychology* 79 (2000): 701–21.

Green, Melanie, and John Donahue. "Simulated Worlds: Transportation into Narratives." In *Handbook of Imagination and Mental Simulation,* edited by Keith Markman, William Klein, and Julie Suhr, 241–54. New York: Psychology Press, 2009.

Green, Melanie, J. Garst, and Timothy Brock. "The Power of Fiction: Determinants and Boundaries." In *The Psychology of Entertainment Media: Blurring the Lines Between Entertainment and Persuasion,* edited by L. J. Shrum. Mahwah, NJ: Erlbaum, 2004.

Greenberg, Daniel. "Flashbulb Memories: How Psychological Research Shows That Our Most Powerful Memories May Be Untrustworthy." *Skeptic,* January 2005, 74–81.

———. "President Bush's False Flashbulb Memory of 9/11." *Applied Cognitive Psychology* 18 (2004): 363–70.

Greitemeyer, Tobias, and Silvia Osswald. "Effects of Prosocial Video Games on Prosocial Behavior." *Journal of Personality and Social Psychology* 98 (2010): 211–21.

Guber, Peter. *Tell to Win: Connect, Persuade, and Triumph with the Hidden Power of Story.* New York: Crown, 2011.

Haidt, Jonathan. *The Happiness Hypothesis.* New York: Basic, 2006.

Hakemulder, Jèmeljan. *The Moral Laboratory: Experiments Examining the Effects of Reading Literature on Social Perception and Moral Self-Concept.* Amsterdam: John Benjamins, 2000.

Hargrove, Thomas. "Third of Americans Suspect 9/11 Government Conspiracy." Scripps News, August 1, 2006. http://www.scrippsnews.com/911poll. Accessed August 30, 2010.

Harris, Paul. *The Work of the Imagination.* New York: Blackwell, 2000.

Haven, Kendall. *Story Proof: The Science Behind the Startling Power of Story.* Westport, CT: Libraries Unlimited, 2007.

Heider, Fritz, and Marianne Simmel. "An Experimental Study of Apparent Behavior." *American Journal of Psychology* 57 (1944): 243–59.

Hemingway, Ernest. *Death in the Afternoon.* New York: Scribner, 1960. First published 1932.

———. *The Old Man and the Sea.* New York: Scribner, 1980. First published 1952.

Hickok, Gregory. "Eight Problems for the Mirror Neuron Theory of Action Understanding in Monkeys and Humans." *Journal of Cognitive Neuroscience* 21 (2009): 1229–43.

Hirstein, William. *Brain Fiction: Self-Deception and the Riddle of Confabulation.* Cambridge, MA: MIT Press, 2006.

———, ed. *Confabulation: Views from Neuroscience, Psychiatry, Psychology, and Philosophy.* Oxford: Oxford University Press, 2009.

Hitchens, Christopher. "The Dark Side of Dickens." *Atlantic Monthly,* May 2010.

Hobson, J. Allan. *Dreaming: An Introduction to the Science of Sleep.* Oxford: Oxford University Press, 2002.

Hood, Bruce. *Supersense: Why We Believe in the Unbelievable.* New York: Harper One, 2009.

Hume, David. "Of the Standard of Taste." *Essays Moral, Practical, and Literary.* London: Longman, Green, and Co., 1875. First published 1757.

Hunt, Harry. "New Multiplicities of Dreaming and REMing." In *Sleep and Dreaming: Scientific Advances and Reconsiderations,* edited by Edward Pace-Schott, Mark Solms, Mark Blagrove, and Stevan Harnad, 164–67. Cambridge, UK: Cambridge University Press, 2003.

Iacoboni, Marco. *Mirroring People: The Science of Empathy and How We Connect with Others.* New York: Picador, 2008.

Icke, David. *The Biggest Secret: The Book That Will Change the World!* Ryde, UK: David Icke Books, 1999.

Jabbi, M., J. Bastiaansen, and C. Keysers. "A Common Anterior Insula Representation of Disgust Observation, Experience and Imagination Shows Divergent Functional Connectivity Pathways." *PLoS ONE* 3 (2008): e2939. doi:10.1371/journal.pone.0002939.

Jacobs, A. J. *The Year of Living Biblically.* New York: Simon and Schuster, 2008.

Jager, Gustav. *The Darwinian Theory and Its Relation to Morality and Religion.* Stuttgart, Germany: Hoffman, 1869.

James, William. *The Will to Believe and Other Essays in Popular Philosophy.* New York: Cosimo, 2007. First published 1897.

Jamison, Kay Redfield. *Touched with Fire: Manic-Depressive Illness and the Artistic Temperament.* New York: Free Press, 1993.

Ji, Daoyun, and Matthew A. Wilson. "Coordinated Memory Replay in the Visual Cortex and Hippocampus During Sleep." *Nature Neuroscience* 10 (2007): 100–107.

Johansson, Petter, Lars Hall, Sverker Sikstrom, and Andreas Olsson. "Failure to Detect Mismatches Between Intention and Outcome in a Simple Decision Task." *Science* 310 (2005): 116–19.

Johnson, Dominic. "Gods of War." In *The Evolution of Religion: Studies, Theories, and Critiques,* edited by Joseph Bulbulia, Richard Sosis, Erica Harris, and Russell Genet. Santa Margarita, CA: Collins Foundation, 2008.

Johnson, Paul. *A History of the American People.* New York: HarperCollins, 1997.

Johnson, Steven. *Everything Bad Is Good for You.* New York: Riverhead, 2005.

Jones, Alex. *Endgame: Blueprint for Global Enslavement.* New York: Disinformation, 2007. Documentary.

Jones, Gerard. *Killing Monsters: Why Children Need Fantasy, Super Heroes, and Make-Believe Violence.* New York: Basic, 2002.

Jouvet, Michael. *The Paradox of Sleep: The Story of Dreaming.* Cambridge, MA: MIT Press, 1999.

Joyce, James. *Finnegans Wake.* New York: Penguin, 1999. First published 1939.

Kagan, Jerome. *The Three Cultures: Natural Sciences, Social Sciences, and the Humanities in the 21st Century.* Cambridge, UK: Cambridge University Press, 2009.

Keen, Suzanne. *Empathy and the Novel.* Oxford: Oxford University Press, 2007.

Kelly, R. V. *Massively Multiplayer Online Role-Playing Games: The People, the Addiction, and the Playing Experience.* Jefferson, NC: McFarland, 2004.

Kershaw, Ian. *Hitler, 1889–1936: Hubris.* New York: Norton, 1998.

Kessel, John. "Invaders." In *The Wesleyan Anthology of Science Fiction,* edited by Arthur B. Evans, Istvan Csicsery-Ronay Jr., Joan Gordon, Veronica Hollinger, Rob Latham, and Carol McGuirk, 654–75. Middletown, CT: Wesleyan University Press, 2010.

Killingsworth, Matthew, and Daniel Gilbert. "A Wandering Mind Is an Unhappy Mind." *Science* 12 (2010): 932.

King, Stephen. *On Writing: A Memoir of the Craft.* New York: Pocket, 2000.

Klinger, Eric. "Daydreaming and Fantasizing: Thought Flow and Motivation." In *Handbook of Imagination and Mental Simulation,* edited by Keith Markman, William Klein, and Julie Suhr, 225–39. New York: Psychology Press, 2009.

Knapp, John, ed. "An Evolutionary Paradigm for Literary Study." Special Issue, *Style* 42/43 (2008).

Koch, Cristof. "Dream States." *Scientific American Mind,* November/December 2010.

Kohler, Joachim. *Wagner's Hitler: The Prophet and His Disciple.* London: Polity, 2000.

Konner, Melvin. *The Evolution of Childhood.* Cambridge, MA: Harvard University Press, 2010.

Krendl, A. C., C. Macrae, W. M. Kelley, J. F. Fugelsang, and T. F. Heatherton. "The Good, the Bad, and the Ugly: An fMRI Investigation of the Functional Anatomic Correlates of Stigma." *Social Neuroscience* 1 (2006): 5–15.

Kubizek, August. *The Young Hitler I Knew.* London: Greenhill, 2006.

Kurzban, Robert. *Why Everyone Else Is a Hypocrite.* Princeton, NJ: Princeton University Press, 2010.

Ledoux, Joseph. *Synaptic Self: How Our Brains Become Who We Are.* New York: Penguin, 2003.

Lehrer, Jonah. *How We Decide.* Boston: Houghton Mifflin, 2009.

Levitin, Daniel J. *The World in Six Songs: How the Musical Brain Created Human Nature.* New York: Plume, 2008.

Lightfoot, Steve. "Who Really Killed John Lennon?: The Truth About His Murder." 2001. http://www.lennonmurdertruth.com/index.asp. Accessed August 30, 2010.

Linn, Susan. *The Case for Make Believe: Saving Play in a Commercialized World.* New York: New Press, 2008.

Loewen, James W. *Lies My Teacher Told Me: Everything Your American History Textbook Got Wrong.* New York: Free Press, 1995.

Loftus, Elizabeth, and Jacqueline Pickrell. "The Formation of False Memories." *Psychiatric Annals* 25 (1995): 720–25.

Ludwig, Arnold. *The Price of Greatness: Resolving the Creativity and Madness Controversy.* New York: Guilford, 1996.

Lynn, Steven Jay, Abigail Matthews, and Sean Barnes. "Hypnosis and Memory: From Bernheim to the Present." In *Handbook of Imagination and Mental Simulation,* edited by Keith Markman, William Klein, and Julie Suhr, 103–18. New York: Psychology Press, 2009.

Maier, Norman. "Reasoning in Humans. II: The Solution of a Problem and Its Appearance in Consciousness." *Journal of Comparative Psychology* 12 (1931): 181–94.

Malcolm, Janet. "Iphigenia in Forest Hills." *The New Yorker,* May 3, 2010.

Mar, Raymond, Maja Djikic, and Keith Oatley. "Effects of Reading on Knowledge,

Social Abilities, and Selfhood." In *Directions in Empirical Literary Studies,* edited by Sonia Zyngier, Marisa Bortolussi, Anna Chesnokova, and Jan Avracher, 127–38. Amsterdam: John Benjamins, 2008.

Mar, Raymond, and Keith Oatley. "The Function of Fiction Is the Abstraction and Simulation of Social Experience." *Perspectives on Psychological Science* 3 (2008): 173–92.

Mar, Raymond, Keith Oatley, Jacob Hirsh, Jennifer dela Paz, and Jordan Peterson. "Bookworms Versus Nerds: Exposure to Fiction Versus Non-Fiction, Divergent Associations with Social Ability, and the Simulations of Fictional Social Worlds." *Journal of Research in Personality* 40 (2006): 694–712.

Mar, Raymond, Keith Oatley, and Jordan Peterson. "Exploring the Link Between Reading Fiction and Empathy: Ruling Out Individual Differences and Examining Outcomes." *Communications: The European Journal of Communication* 34 (2009): 407–28.

Marcus, Gary. *Kluge: The Haphazard Evolution of the Human Mind.* Boston: Mariner, 2008.

Marsh, Elizabeth, Michelle Meade, and Henry Roediger III. "Learning Facts from Fiction." *Journal of Memory and Language* 49 (2003): 519–36.

Mastro, Dana. "Effects of Racial and Ethnic Stereotyping." In *Media Effects: Advances in Theory and Research,* 3rd ed., edited by Jennings Bryant and Mary Beth Oliver. New York: Routledge, 2009.

Maugham, Somerset. *Ten Novels and Their Authors.* New York: Penguin, 1969.

McAdams, Dan. "Personal Narratives and the Life Story." In *Handbook of Personality: Theory and Research,* edited by Oliver John, Richard Robins, and Lawrence Pervin, 241–61. New York: Guilford, 2008.

———. "The Psychology of Life Stories." *Review of General Psychology* 5 (2001): 100–122.

———. *The Stories We Live By: Personal Myths and the Making of the Self.* New York: Guilford, 1993.

McGonigal, Jane. *Reality Is Broken: Why Games Make Us Better and How They Can Change the World.* New York: Penguin, 2011.

McLuhan, Marshall. *The Gutenberg Galaxy: The Making of Typographic Man.* Toronto: University of Toronto Press, 1962.

McNamara, Patrick. *An Evolutionary Psychology of Sleep and Dreams.* Westport, CT: Praeger, 2004.

Meadows, Mark Stephen. *I, Avatar: The Culture and Consequences of Having a Second Life.* Berkeley, CA: New Rider, 2008.

Mele, Alfred. *Self-Deception Unmasked.* Princeton, NJ: Princeton University Press, 2001.

Meltzoff, Andrew, and Jean Decety. "What Imitation Tells Us About Social Cognition: A Rapprochement Between Developmental Psychology and Cognitive Neuroscience." *Philosophical Transactions of the Royal Society, London B* 358 (2003): 491–500.

Meltzoff, Andrew, and M. Keith Moore. "Imitation of Facial and Manual Gestures by Human Neonates." *Science* 198 (1977): 75–78.

Metaxas, Eric. *Bonhoeffer: Pastor, Martyr, Prophet, Spy.* Nashville: Thomas Nelson, 2010.

Miller, Geoffrey. *The Mating Mind.* New York: Anchor, 2001.

Miller, Laura. "The Last Word: How Many Books Are Too Many?" *New York Times,* July 18, 2004.

Miller, Rory. *Meditations on Violence.* Wolfeboro, NH: YMAA Publication Center, 2008.

Morley, Christopher. *Parnassus on Wheels.* New York: Doubleday, 1917.

Motion Picture Association of America Worldwide Market Research and Analysis. *U.S. Entertainment Industry: 2006 Market Statistics.* http://www.google.com/#sclient=psy&hl=en&source=hp&q=US+Entertainment+Industry:+2006+Market+Statistics&aq=f&aqi=&aql=&oq=&pbx=1&bav=on.2,or.r_gc.r_pw.&fp=ca5f50573a0b59e3&biw=1024&bih=571. Accessed July 30, 2011.

Nabokov, Vladimir. *Pale Fire.* New York: Vintage, 1989. First published 1962.

Napier, John. *Hands.* Princeton, NJ: Princeton University Press, 1993.

National Endowment for the Arts. *Reading on the Rise: A New Chapter in American Literacy.* 2008. http://www.nea.gov/research/ReadingonRise.pdf. Accessed August 30, 2010.

Neisser, Ulric, and Nicole Harsch. "Phantom Flashbulbs: False Recollections of Hearing the News About Challenger." In *Affect and Accuracy in Recall: Studies of "Flashbulb" Memories,* vol. 4, edited by Eugene Winograd and Ulric Neisser, 9–31. Cambridge, UK: Cambridge University Press, 1992.

Nell, Victor. *Lost in a Book: The Psychology of Reading for Pleasure.* New Haven, CT: Yale University Press, 1988.

Nettle, Daniel. *Strong Imagination: Madness, Creativity, and Human Nature.* Oxford: Oxford University Press, 2001.

Nicholson, Christopher. *Richard and Adolf.* Jerusalem: Gefen, 2007.

Niles, John. *Homo Narrans: The Poetics and Anthropology of Oral Literature.* Philadelphia: University of Pennsylvania Press, 1999.

Norrick, Neal. "Conversational Storytelling." In *The Cambridge Companion to Narrative,* edited by David Herman, 127–41. Cambridge, UK: Cambridge University Press, 2007.

Oatley, Keith. "The Mind's Flight Simulator." *Psychologist* 21 (2008): 1030–32.

———. "The Science of Fiction." *New Scientist,* June 25, 2008.

———. *Such Stuff as Dreams: The Psychology of Fiction.* New York: Wiley, 2011.

Olmsted, Kathryn. *Real Enemies.* Oxford: Oxford University Press, 2009.

Ong, Walter. *Orality and Literacy.* New York: Routledge, 1982.

Ost, James, Granhag Pär-Anders, Julie Udell, and Emma Roos af Hjelmsäter. "Familiarity Breeds Distortion: The Effects of Media Exposure on False Reports Concerning Media Coverage of the Terrorist Attacks in London on 7 July 2005." *Memory* 16 (2008): 76–85.

Paley, Vivian. *Bad Guys Don't Have Birthdays: Fantasy Play at Four.* Chicago: University of Chicago Press, 1988.

———. *Boys and Girls: Superheroes in the Doll Corner.* Chicago: University of Chicago Press, 1984.

———. *A Child's Work: The Importance of Fantasy Play.* Chicago: University of Chicago Press, 2004.

Philbrick, Nathaniel. *In the Heart of the Sea.* New York: Penguin, 2000.

Pinker, Steven. *The Better Angels of Our Nature: Why Violence Has Declined.* New York: Viking, 2011.

———. *The Blank Slate.* New York: Viking, 2002.

———. *How the Mind Works.* New York: Norton, 1997.

———. "Toward a Consilient Study of Literature." *Philosophy and Literature* 31 (2007): 161–77.

Pinsky, Robert. *The Handbook of Heartbreak: 101 Poems of Lost Love and Sorrow.* New York: Morrow, 1998.

Plato. *The Republic.* Translated by Desmond Lee. New York: Penguin, 2003.

Poe, Edgar Allan. *Complete Tales and Poems of Edgar Allan Poe.* New York: Vintage, 1975.

Pronin, Emily, Daniel Linn, and Lee Ross. "The Bias Blind Spot: Perceptions of Bias in Self Versus Others." *Personality and Social Psychology Bulletin* 28 (2002): 369–81.

Ramachandran, V. S. *The Tell-Tale Brain: A Neuroscientist's Quest for What Makes Us Human.* New York: Norton, 2011.

Reeves, Byron, and Clifford Nass. *The Media Equation: How People Treat Computers, Television, and New Media Like Real People and Places.* Stanford, CA: CSLI Publications, 2003.

Revonsuo, Antti. "Did Ancestral Humans Dream for Their Lives?" In *Sleep and Dreaming: Scientific Advances and Reconsiderations,* edited by Edward Pace-Schott, Mark Solms, Mark Blagrove, and Stevan Harnad, 275–94. Cambridge: Cambridge University Press, 2003.

———. "The Reinterpretation of Dreams: An Evolutionary Hypothesis of the Function of Dreaming." *Behavioral and Brain Sciences* 23 (2000): 793–1121.

Richardson, Samuel. *The History of Sir Charles Grandison.* Vol. 6. London: Suttaby, Evance and Fox, 1812. First published 1753–1754.

Rizzolatti, Giacomo, Corrando Sinigaglia, and Frances Anderson. *Mirrors in the Brain: How Our Minds Share Actions, Emotions, and Experiences.* Oxford: Oxford University Press, 2008.

Rock, Andrea. *The Mind at Night: The New Science of How and Why We Dream.* New York: Basic, 2004.

Roskos-Ewoldsen, David, Beverly Roskos-Ewoldsen, and Francesca Carpentier. "Media Priming: An Updated Synthesis." *In Media Effects: Advances in Theory and Research,* 3rd ed., edited by Jennings Bryant and Mary Beth Oliver. New York: Routledge, 2009.

Russell, David. *Literature for Children: A Short Introduction.* 2nd ed. New York: Longman, 1991.

Sacks, Oliver. *The Man Who Mistook His Wife for a Hat.* New York: Simon and Schuster, 1985. First published 1970.

Schachter, Daniel. *Searching for Memory: The Brain, the Mind, and the Past.* New York: Basic, 1996.

——. *The Seven Sins of Memory: How the Mind Forgets and Remembers.* Boston: Houghton Mifflin, 2001.

Schechter, Harold. *Savage Pastimes: A Cultural History of Violent Entertainment.* New York: St. Martin's, 2005.

Schweikart, Larry, and Michael Allen. *A Patriot's History of the United States: From Columbus's Great Discovery to the War on Terror.* New York: Sentinel, 2007.

Shaffer, David, S. A. Hensch, and Katherine Kipp. *Developmental Psychology.* New York: Wadsworth, 2006.

Shedler, Jonathan. "The Efficacy of Psychodynamic Therapy." *American Psychologist* 65 (2010): 98–109.

Shields, David. *Reality Hunger: A Manifesto.* New York: Knopf, 2010.

Shirer, William L. *The Rise and Fall of the Third Reich.* New York: Simon and Schuster, 1990.

Singer, Dorothy, and Jerome Singer. *The House of Make Believe: Play and the Developing Imagination.* Cambridge, MA: Harvard University Press, 1990.

Slater, Mel, Angus Antley, Adam Davison, David Swapp, Christoph Guger, Chris Barker, Nancy Pistrang, and Maria V. Sanchez-Vives. "A Virtual Reprise of the Stanley Milgram Obedience Experiments." *PLoS ONE* 1 (2006).

Smith, Stacy, and Amy Granados. "Content Patterns and Effects Surrounding Sex-Role Stereotyping on Television and Film." In *Media Effects: Advances in Theory and Research,* 3rd ed., edited by Jennings Bryant and Mary Beth Oliver. New York: Routledge, 2009.

Smoking Gun. "A Million Little Lies," January 8, 2006. http://www.thesmokinggun.com/documents/celebrity/million-little-lies. Accessed August 30, 2010.

Solms, Mark. "Dreaming and REM Sleep Are Controlled by Different Mechanisms." In *Sleep and Dreaming: Scientific Advances and Reconsiderations,* edited by Edward Pace-Schott, Mark Solms, Mark Blagrove, and Stevan Harnad. Cambridge: Cambridge University Press, 2003.

Speer, Nicole, Jeremy Reynolds, Khena Swallow, and Jeffrey M. Zacks. "Reading Stories Activates Neural Representations of Visual and Motor Experiences." *Psychological Science* 20 (2009): 989–99.

Spence, Donald. *Narrative Truth and Historical Truth and the Freudian Metaphor.* New York: Norton, 1984.

Spotts, Frederic. *Hitler and the Power of Aesthetics.* New York: Overlook, 2003.

Stone, Jason. "The Attraction of Religion." In *The Evolution of Religion: Studies, Theories, and Critiques,* edited by Joseph Bulbulia, Richard Sosis, Erica Harris, and Russell Genet. Santa Margarita, CA: Collins Foundation, 2008.

Stowe, Charles, and Lyman Beecher Stowe. *Harriet Beecher Stowe: The Story of Her Life.* Boston: Houghton Mifflin, 1911.

Stowe, Harriet Beecher. *Uncle Tom's Cabin.* New York: Norton, 2007. First published 1852.

Sugiyama, Michelle Scalise. "Reverse-Engineering Narrative: Evidence of Special Design." In *The Literary Animal,* edited by Jonathan Gottschall and David Sloan Wilson. Evanston, IL: Northwestern University Press, 2005.

Sutton-Smith, Brian. *The Ambiguity of Play.* Cambridge, MA: Harvard University Press, 1997.

——. "Children's Fiction Making." In *Narrative Psychology: The Storied Nature of Human Conduct,* edited by Theodore Sarbin. New York: Praeger, 1986.

Swift, Graham. *Waterland.* New York: Penguin, 2010. First published 1983.

Talbot, Margaret. "Nightmare Scenario." *The New Yorker,* November 16, 2009.

Taleb, Nassim. *The Black Swan: The Impact of the Highly Improbable.* New York: Penguin, 2008.

Tallis, Raymond. *The Hand: A Philosophical Inquiry into Human Being.* Edinburgh: Edinburgh University Press, 2003.

Tanaka, Jiro. "What Is Copernican? A Few Common Barriers to Darwinian Thinking About the Mind." *Evolutionary Review* 1 (2010): 6–12.

Tatar, Maria. *The Hard Facts of the Grimms' Fairy Tales.* 2nd ed. Princeton, NJ: Princeton University Press, 2003.

Tavris, Carol, and Eliot Aronson. *Mistakes Were Made but Not by Me: Why We Justify Foolish Beliefs, Bad Decisions, and Hurtful Acts.* New York: Harcourt, 2007.

Taylor, Marjorie. *Imaginary Companions and the Children Who Create Them.* Oxford: Oxford University Press, 1999.

Taylor, Shelley. *Positive Illusions: Creative Self-Deception and the Healthy Mind.* New York: Basic, 1989.

Tolstoy, Leo. *What Is Art?* New York: Crowell, 1899.

Tsai, Michelle. "Smile and Say 'Fat!'" *Slate,* February 22, 2007. http://www.slate.com/id/2160377/. Accessed March 4, 2011.

The Ultimate Fighter. Spike TV. Episode 11, season 10, December 2, 2009.

United States Holocaust Memorial Museum. "Book Burning." In *Holocaust Encyclo-*

pedia. http://www.ushmm.org/wlc/en/article.php?ModuleId=10005852. Accessed August 30, 2010.

Valli, Katja, and Antti Revonsuo. "The Threat Simulation Theory in Light of Recent Empirical Evidence: A Review." *American Journal of Psychology* 122 (2009): 17–38.

Viereck, Peter. *Metapolitics: The Roots of the Nazi Mind.* New York: Capricorn, 1981. First published 1941.

Voland, Eckart, and Wulf Schiefenhovel, eds. *The Biological Evolution of Religious Mind and Behavior.* Dordrecht, Germany: Springer, 2009.

Wade, Nicholas. *The Faith Instinct: How Religion Evolved and Why It Endures.* New York: Penguin, 2009.

Waller, Douglas. *Air Warriors: The Inside Story of the Making of a Navy Pilot.* New York: Dell, 1999.

Wallis, James. "Making Games That Make Stories." In *Second Person: Role-Playing and Story in Games and Playable Media,* edited by Pat Harrigan and Noah Wardrip-Fruin. Cambridge, MA: MIT Press, 2007.

Walton, Kendall. *Mimesis as Make Believe: On the Foundations of the Representational Arts.* Cambridge, MA: Harvard University Press, 1990.

Weinstein, Cindy. *Cambridge Companion to Harriet Beecher Stowe.* Cambridge, UK: Cambridge University Press, 2004.

Weisberg, Deena Skolnick. "The Vital Importance of Imagination." In *What's Next: Dispatches on the Future of Science,* edited by Max Brockman. New York: Vintage, 2009.

Wheatley, Thalia. "Everyday Confabulation." In *Confabulation: Views from Neuroscience, Psychiatry, Psychology, and Philosophy,* edited by William Hirstein, 203–21. Oxford: Oxford University Press, 2009.

Wiesel, Elie. *The Gates of the Forest.* New York: Schocken, 1966.

Wilson, Anne, and Michael Ross. "From Chump to Champ: People's Appraisals of Their Earlier and Present Selves." *Journal of Personality and Social Psychology* 80 (2000): 572–84.

Wilson, David Sloan. *Darwin's Cathedral.* Chicago: University of Chicago Press, 2003.

———. "Evolution and Religion: The Transformation of the Obvious." In *The Evolution of Religion: Studies, Theories, and Critiques,* edited by Joseph Bulbulia, Richard Sosis, Erica Harris, and Russell Genet. Santa Margarita, CA: Collins Foundation, 2008.

———. *Evolution for Everyone.* New York: Random House, 2007.

Wilson, David Sloan, and Edward O. Wilson. "Rethinking the Theoretical Foundation of Sociobiology." *Quarterly Review of Biology* 82 (2007): 327–48.

Wilson, Edward O. *Consilience: The Unity of Knowledge.* New York: Knopf, 1998.

Wilson, Frank. *The Hand: How Its Use Shapes the Brain, Language, and Human Culture.* New York: Pantheon, 1998.

Wolfe, Tom. "The New Journalism." In *The New Journalism,* edited by Tom Wolfe and Edward Warren Johnson, 13–68. London: Picador, 1975.

Wood, James. *How Fiction Works.* New York: Picador, 2008.

Wood, Wendy, and Alice Eagly. "A Cross-Cultural Analysis of the Behavior of Women and Men: Implications for the Origins of Sex Differences." *Psychological Bulletin* 128 (2002): 699–727.

Yagoda, Ben. *Memoir: A History.* New York: Riverhead, 2009.

Young, Kay, and Jeffrey Saver. "The Neurology of Narrative." *Substance* 94/95 (2001): 72–84.

Zinn, Howard. *A People's History of the United States, 1492–Present.* New York: Harper, 2003. First published 1980.

Zunshine, Lisa. *Why We Read Fiction.* Columbus: Ohio University Press, 2006.

Credits

page xii: Vintage Images/Alamy. page 2: Bettman/Corbis. page 6: Jonathan Gottschall. page 7: © English Heritage/NMR. page 9: © Aaron Escobar. page 13: Airman 1st Class Nicholas Pilch. page 17: PASIEKA/SPL/Getty Images. page 19: Nat Farbman/Getty Images. page 24: © Tonny Tunya, Compassion International, used by permission. page 25: © Charles and Josette Lenars/Corbis. page 26: Photograph by Bailey Rae Weaver, whose photography can be viewed at www.flickr.com/photos/baileysjunk. page 31: © Peter Turnley/Corbis. page 37: Corbis. page 40: Corbis. page 43: From *More English Fairy Tales,* ed. Joseph Jacobs, illus. John D. Batten, G. Putnam's Sons, 1922. page 51: From *George Gissing: A Critical Study,* Frank Swinnerton, Martin Secker, 1912. page 53: Dorothea Lange/Library of Congress. page 57: Corbis. page 58: David Shankbone. page 60: From Andrew N. Meltzoff and M. Keith Moore, "Imitation of Facial and Manual Gestures by Human Neonates." *Science,* 198, 1977, 75–78. page 63: Public Library of Science. page 65: From *The History of Don Quixote,* Miguel de Cervantes, illus. Gustave Doré, Cassell and Co., 1906. page 69: Photo courtesy of D. Sharon Pruitt. page 70: Corbis. page 72: Corbis. page 74: Corbis. page 78: Getty Images. page 81: Nicky Wilkes, Redditch, UK. page 90: *The Air Loom, A Human Influencing Machine,* 2002, Rod Dickinson. page 94: Pinguino Kolb. page 97: *The Accidental Mind: How Brain Evolution Has Given Us Love, Memory, Dreams, and God* by David J. Linden, p. 228, Cambridge, Mass.: The Belknap Press of Harvard University Press, Copyright © 2007 by the President and Fellows of Harvard College. Joan M. K. Tycko, illustrator. page 100: From *Tales of Sherlock Holmes,* A. Conan Doyle, A. L. Burt, 1906. page 103: Courtesy NASA/JPL-Caltech. page 105: Jonathan Gottschall. page 107: Corbis, Corbis, Photodisc/Getty Images, Corbis. page 109: Python (Monty) Pictures. page 112: Roman Suzuki. page 114: smokinggun.com. page 118: Getty Images. page 123: Library of Congress. page 126: Corbis. page 131: From *Madame Bovary: A Tale of Provincial Life,* Gustave Flaubert, M. Walter Dunne, 1904. page 133: From *Oriental Cairo,* Douglas Sladen,

Index

page references in italics refer to text graphics